Lycopene and Tomatoes in Human Nutrition and Health

Lycopene and Tomatoes in Human Nutrition and Health

Edited by
A. Venketeshwer Rao
Gwen L. Young
Leticia G. Rao

CRC Press
Taylor & Francis Group
Boca Raton London New York

CRC Press is an imprint of the
Taylor & Francis Group, an **informa** business

CRC Press
Taylor & Francis Group
6000 Broken Sound Parkway NW, Suite 300
Boca Raton, FL 33487-2742

First issued in paperback 2021

© 2018 by Taylor & Francis Group, LLC
CRC Press is an imprint of Taylor & Francis Group, an Informa business

No claim to original U.S. Government works

ISBN 13: 978-1-03-209564-6 (pbk)
ISBN 13: 978-1-4665-7537-0 (hbk)

This book contains information obtained from authentic and highly regarded sources. Reasonable efforts have been made to publish reliable data and information, but the author and publisher cannot assume responsibility for the validity of all materials or the consequences of their use. The authors and publishers have attempted to trace the copyright holders of all material reproduced in this publication and apologize to copyright holders if permission to publish in this form has not been obtained. If any copyright material has not been acknowledged please write and let us know so we may rectify in any future reprint.

Except as permitted under U.S. Copyright Law, no part of this book may be reprinted, reproduced, transmitted, or utilized in any form by any electronic, mechanical, or other means, now known or hereafter invented, including photocopying, microfilming, and recording, or in any information storage or retrieval system, without written permission from the publishers.

For permission to photocopy or use material electronically from this work, please access www.copyright.com (http://www.copyright.com/) or contact the Copyright Clearance Center, Inc. (CCC), 222 Rosewood Drive, Danvers, MA 01923, 978-750-8400. CCC is a not-for-profit organization that provides licenses and registration for a variety of users. For organizations that have been granted a photocopy license by the CCC, a separate system of payment has been arranged.

Trademark Notice: Product or corporate names may be trademarks or registered trademarks, and are used only for identification and explanation without intent to infringe.

Library of Congress Cataloging-in-Publication Data

Names: Rao, A. Venket, editor.
Title: Lycopene and tomatoes in human nutrition and health / edited by A. Venketeshwer Rao, Gwen L. Young and Leticia Rao.
Description: Boca Raton : Taylor & Francis, 2018. | Includes bibliographical references and index.
Identifiers: LCCN 2018005768 | ISBN 9781466575370 (hardback : alk. paper)
Subjects: LCSH: Lycopene--Health aspects. | Tomatoes--Nutrition.
Classification: LCC QK898.L9 L9235 2018 | DDC 572/.592--dc23
LC record available at https://lccn.loc.gov/2018005768

Visit the Taylor & Francis Web site at
http://www.taylorandfrancis.com

and the CRC Press Web site at
http://www.crcpress.com

Publisher's Note
The publisher has gone to great lengths to ensure the quality of this reprint but points out that some imperfections in the original copies may be apparent.

Contents

Preface

There is compelling scientific evidence based on in vitro animal models, epidemiological and preclinical studies, and clinical studies to indicate that oxidative stress due to the accumulation of free radicals such as the reactive oxygen species (ROS) is related to increased risk of several, if not all, human diseases. Antioxidants—by virtue of their preferential interactions with ROS—can and have been shown in many studies to mitigate the damaging effect of ROS on cellular proteins, lipids, and DNA. The imbalance between ROS and antioxidants that can shift in favor of free radical accumulation is of major concern. It is reasonable, therefore, to assume that consuming diets rich in antioxidants as well as antioxidants in the form of nutritional supplements may play an important role in reducing the risk of human diseases. Indeed, several epidemiological studies have supported such an assumption. Diets rich in fruit and vegetables that are good sources of antioxidants have indeed been shown to lower several human chronic diseases. Antioxidants include a wide spectrum of compounds from vitamins such as A, C, and E to biologically active phytonutrients, which include the class of compounds known as carotenoids and polyphenols. Undoubtedly, fruit and vegetables also contain many other compounds with antioxidant properties. Lycopene and β-carotene are the two major carotenoids present in foods as well as in circulation, tissues, and organs. Lycopene, in particular, is known as the most potent of all natural antioxidants, attracting scientific interest in its role in human health and disease. Initially, it was organic chemists who provided information about its chemical structure as well as its antioxidant properties. Almost three decades later, scientists started studying the metabolism, bioavailability, and tissue accumulation of lycopene. In 1995, an epidemiological study showed that diets rich in lycopene lowered the risk of prostate cancer. Followed by this groundbreaking observation, scientists began studying the scope of lycopene in other cancers as well as coronary arterial disease (CVD), with encouraging results. However, it is now evident that human diseases other than cancers and CVD are also very common around the globe and contribute to high levels of mortality and morbidity. The knowledge that other human diseases may also be related to oxidative stress motivated investigations into the protective role of lycopene and tomatoes in diabetes, bone diseases such as osteoporosis, respiratory disorders, hypertension, male infertility, inflammatory diseases such as rheumatoid arthritis, neurodegenerative diseases, skin diseases, and ocular diseases, to name a few.

Over the years several books were published addressing the topic of lycopene, tomatoes, and human nutrition and health. The last book to update our knowledge was almost a decade ago. This provided us with an opportunity to publish a book that not only looks at the historical development of lycopene in relation to its occurrence, chemistry, bioavailability, and its role in the prevention of diseases, but one that can bring the knowledge up to date. We were very fortunate that several world-renowned scientists with expertise in various disciplines agreed to contribute to the book. The topics of the book chapters range from newer information on tomato phytonutrients, lycopene chemistry, metabolism, and bioavailability to their role in a number of major and important diseases. We also have industrial and regulatory representation among our authors that share relevant information from health claims to food processing opportunities to maximize bioactives and increase food nutrition. This book should serve as a reference book not only for researches in this area but also for many other stake holders, including public health professionals, regulators, and the food, nutritional supplement, and pharmaceutical industries, as well as the general public at large who are interested in diet, nutrition, and health. It is our hope that this book will provide not only further understanding of the roles of lycopene and tomatoes in preventing diet-related diseases, but ultimately an improvement in global human health.

We offer our sincerest thanks to all the contributing authors for their excellent coverage of the information in their respective areas of expertise and wish them great successes with their ongoing studies. Thanks also go to the publisher for all the patience and help that they provided and for making this book a reality.

Editors

Dr. A. Venketeshwer Rao, Professor Emeritus, Department of Nutritional Sciences, Faculty of Medicine, University of Toronto, has established a major focus in the area of diet and health. His research has focused on the role of oxidative stress and antioxidant phytochemicals in the causation and prevention of chronic diseases, with particular emphasis on the role of carotenoids and polyphenols. He is one of the pioneering researchers to study the bioavailability, metabolism, mechanisms of action, and biological role of lycopene, a carotenoid antioxidant present in tomatoes, tomato products, other fruits and vegetables, and nutritional supplements, and is credited for bringing international awareness to the role of lycopene in human health. His research interests also include the role of prebiotics and probiotics in human health, and he has over 100 publications in scientific journals and several books and book chapters. Dr. Rao's distinguished academic career spans over 52 years, and he is popularly sought ought by the international media to express his opinions on the subjects of nutrition and health.

Dr. Leticia G. Rao is Adjunct Professor of Medicine and a full member of the Graduate Faculty, Institute of Medical Sciences, at the University of Toronto. She is the Director of the Calcium Research Laboratory and an Emeritus Scientist at St. Michael's Hospital, Toronto, Ontario. Her postgraduate degrees include an M.Sc. in Food Science from Oregon State University and a Ph.D. in Biochemistry from the University of Toronto, and her expertise is in the area of bone cell biology and calcium metabolism, with the long-term goal of uncovering the causes and prevention of osteoporosis by studying bone cells in the laboratory. She has extensive basic research experience on drugs that have been approved for osteoporosis prevention and treatment, such as raloxifene and estrogen. She has conducted clinical studies on postmenopausal women to delineate how the fat-soluble carotenoid antioxidant lycopene present predominantly in tomatoes and tomato products and the water-soluble antioxidant polyphenols in natural food supplements can help in the prevention of risk for postmenopausal osteoporosis. Dr. Rao has presented her research data and has been invited to give talks on her clinical research on osteoporosis to a number of international symposia, and her publications in peer-reviewed scientific journals are extensive. She has co-authored

a book entitled *Bone Building Solution* and co-edited the books *Phytochemicals –
Isolation, Characterization and Role in Human Health* and *Probiotics and Prebiotics
in Human Nutrition and Health.* Although presently retired, she still holds academic
positions both at the University of Toronto and St. Michael's Hospital, and actively
presents her data to International Scientific Symposia.

Gwen L. Young is a world-recognized leader in advo-
cating and promoting tomato and tomato product
academic research, as well as science-based healthy
foods, diets, and access for all.

She is president of the Tomato Foundation, where
she educates and inspires lifestyle as medicine and
aims to reduce the incidence of diet-related diseases and promote access to healthy plant-based
foods. She is also the chair of the Tomato and Health
Commission at the World Processing Tomato Council
and is responsible for the analysis and promotion of
scientifically valid global academic research pub-
lished on the health benefits of tomatoes and tomato products. Additionally, as a member of the science committee for ISHS she has provided scientists worldwide with
opportunities to present and publish their research and obtain research funding. Gwen
completing her BS and MS degrees in Food Biochemistry/Food Science at UC Davis
and also spent the next 25 years directing product development, marketing, culinary
arts, quality assurance, and scientific advising for Nestle Brands and Kagome Inc.

Contributors

Koichi Aizawa
Innovation Division
Kagome Co, Ltd
Nasushiobara, Tochigi, Japan

Patrick Borel
Center of Cardio Vascular and Nutrition
    research (C2VN)
Faculty of Medicine
Aix-Marseille University
Marseille, France

Volker Böhm
Institute of Nutrition
Friedrich Schiller University Jena
Jena, Germany

Montaña Cámara
ALIMNOVA Research Group
Nutrition and Food Science Department
Pharmacy Faculty
University Complutense of Madrid
Madrid, Spain

Rosa María Cámara
ALIMNOVA Research Group
Nutrition and Food Science Department
Pharmacy Faculty
University Complutense of Madrid
Madrid, Spain

Charles Desmarchelier
Center of Cardio Vascular and Nutrition
    research (C2VN)
Faculty of Medicine
Aix-Marseille University
Marseille, France

Laura Domínguez
ALIMNOVA Research Group
Nutrition and Food Science Department
Pharmacy Faculty
University Complutense of Madrid
Madrid, Spain

Soumia Fenni
Center of Cardio Vascular and Nutrition
    research (C2VN)
Faculty of Medicine
Aix-Marseille University
Marseille, France

Virginia Fernández-Ruiz
ALIMNOVA Research Group
Nutrition and Food Science Department
Pharmacy Faculty
University Complutense of Madrid
Madrid, Spain

Daniel L. Graham
School of Natural Sciences and
    Psychology
Liverpool John Moores University
Liverpool, UK

Takuro Inoue
Innovation Division
Kagome Co, Ltd
Nasushiobara, Tochigi, Japan

Teruaki Iwamoto
Center for Infertility and IVF
International University of Health and
    Welfare Hospital
Nasushiobara, Tochigi, Japan

Jean-François Landrier
Center of Cardio Vascular and Nutrition
    research (C2VN)
Faculty of Medicine
Aix-Marseille University
Marseille, France

Gordon M. Lowe
School of Pharmacy and Biomolecular
    Sciences
Liverpool John Moores University
Liverpool, UK

Leticia G. Rao
University of Toronto
St. Michael's Hospital
Toronto, Ontario, Canada

A. Venketeshwer Rao
Department of Nutritional Sciences
Faculty of Medicine
University of Toronto
Toronto, Ontario, Canada

Luca Sandei
Department of Tomato Products
SSICA Foundation of Research
Stazione Sperimentale per l'Industria
    Conserve Alimentari
Parma, Italy

Ikuo Sato
Department of Obstetrics and
    Gynecology
International University of Health and
    Welfare Hospital
Nasushiobara, Tochigi, Japan

M. Cortes Sánchez-Mata
ALIMNOVA Research Group
Nutrition and Food Science Department
Pharmacy Faculty
University Complutense of Madrid
Madrid, Spain

Hiroyuki Suganuma
Innovation Division
Kagome Co, Ltd
Nasushiobara, Tochigi, Japan

Franck Tourniaire
Center of Cardio Vascular and Nutrition
    research (C2VN)
Faculty of Medicine
Aix-Marseille University
Marseille, France

Yu Yamamoto
Innovation Division
Kagome Co, Ltd
Nasushiobara, Tochigi, Japan

Andrew J. Young
School of Natural Sciences and
    Psychology
Liverpool John Moores University
Liverpool, UK

Gwen L. Young
The Tomato Foundation
California

1 Lycopene
Chemistry, Metabolism, and Bioavailability

Gordon M. Lowe, Daniel L. Graham, and Andrew J. Young

CONTENTS

1.1 CHEMISTRY

Lycopene (ψ,ψ–carotene) is best known as the main carotenoid from the red tomato fruit (*Lycopersicum esculentum*), but it also found in some other fruits (e.g., watermelon). A closely related compound, the tetra-(Z)-isomer prolycopene (7Z,9Z,7′Z,9′Z–ψ,ψ–carotene) is the main carotenoid of tangerine tomato mutants (e.g., var. Tangella and var. Ailsa Craig). Lycopene possesses a simple hydrocarbon structure ($C_{40}H_{56}$) characterized by its 11 conjugated double bonds as a polyene chain (Figure 1.1). This polyene chain primarily influences the chemistry (e.g., redox properties), location, and orientation of lycopene within lipid bilayers (see below). Combined, these influence the behaviour of lycopene and related compounds in biological systems.

Whilst the all-*E* form of lycopene predominates in foods, several geometric isomers of lycopene have been isolated from natural sources or semi-synthesized (typically by iodine- or heat-catalization). Details for the synthesis of geometric isomers of lycopene, together with their chromatographic and spectroscopic characterization, have been published (Hengartner et al. 1992). The structures of the major Z-isomers of lycopene found in biological systems are shown in Figure 1.1. In addition to the all-*E* form in raw and cooked tomatoes and their products, the main isomers are

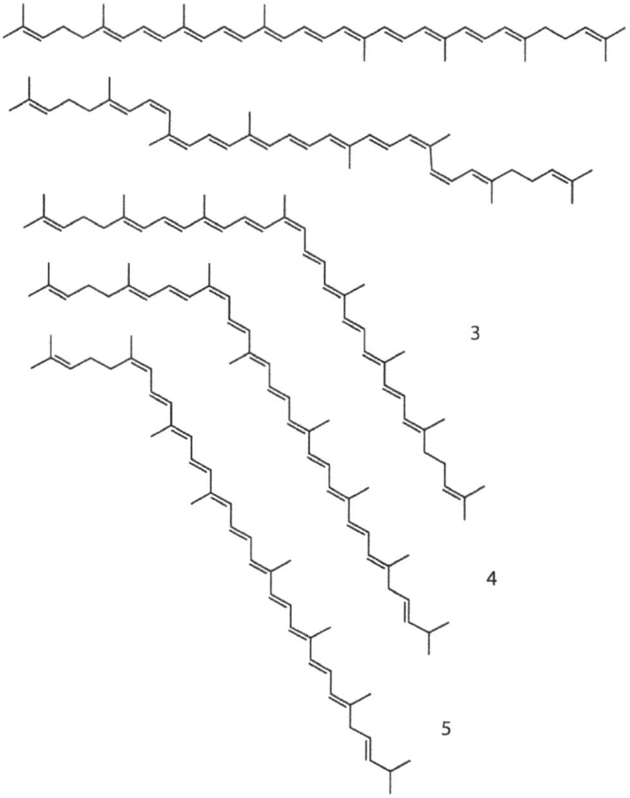

FIGURE 1.1 Structures of the most commonly found geometric forms of lycopene and pro-lycopene. (1) All-*E*-lycopene; (2) prolycopene; (3) 13-Z-lycopene; (4) 9-Z-lycopene; (5) 5-Z-lycopene.

the 5Z, 9Z, 13Z, and 15Z. The proportion of different isomeric forms of lycopene in foods and foodstuffs varies. In fruits, the all-*E* form typically accounts for the vast majority of lycopene (up to 97%), whilst in cooked and processed products such as tomato paste this may fall to 35% (Schierle et al. 1997).

This is of importance because the physico-chemical properties of these different isomeric forms of lycopene vary. Some of the properties of the main geometric isomers are shown in Table 1.1. An *in silico* study (Chasse et al. 2001) indicated that the 5Z and all-*E* isomers of lycopene were the most energetically stable forms of the molecule. Indeed, the 5Z form was found to be the most stable form of the carotenoid, in contrast to the results of a previous study (Berg et al. 2000). Chasse et al. (2001) also reported that all-*E* lycopene is not planar in form. The various Z-isomers of lycopene are recognised as having better solubility than the All-*E* form due to a higher steric demand. In some studies, it is possible that the higher-reported

TABLE 1.1
Spectral Characteristics of the Major Geometric Isomers of Lycopene

A. Spectral characteristics of the major geometric isomers of lycopene in common solvents

Isomer	λ_{max} (nm)	Absorption Coefficient $\left(E_{1cm}^{1\%}\right)$	Solvent	Reference
all-E-lycopene	472	3450	n-hexane, 2% CH$_2$Cl$_2$	Schierle et al. (1997)
5Z-lycopene	470	3466	n-hexane, 2% CH$_2$Cl$_2$	Aebischer et al. (1999)
9Z-lycopene	464	3241	n-hexane	Müller et al. (1997)
13Z-lycopene	463	2533	n-hexane, 2% CH$_2$Cl$_2$	Müller et al. (1997)
7Z,9Z,7'Z,9'Z-lycopene	437	1956	n-hexane, 2% CH$_2$Cl$_2$	Hengartner et al. (1992)

B. On-line spectral characteristics determined by diode-array detection (in the range 200–600 nm) using HPLC separation by a C30 reversed-phase column[a]

Isomer	λ_{max} (nm)				Ab/All	T$_R$ (min)
all-E-lycopene		446	472	503	–	29.5
5Z-lycopene		446	472	503	–	32.0
9Z-lycopene	362	440	466	498	0.13	20.7
13Z-lycopene	360	442	466	498	0.58	16.3

[a] See Graham et al. 2010 for details.

antioxidant activity of the Z-isomers could be, at least in part, due to this improved solubility coupled with a lower tendency to self-aggregate (crystallize) in polar media. Simple hydrocarbon carotenoids such as lycopene display a strong tendency to (micro)-crystallize and some of the reported reactions with radicals may be, at least in part, influenced by such aggregation (El-Agamey et al. 2004), especially where high concentrations of lycopene are present. Such behaviour is more evident in acyclic molecules such as lycopene compared to cyclic carotenoids such as β-carotene (β,β-carotene). In tomatoes, the high concentrations of lycopene naturally occur in crystalline form in the chromoplasts of the fruit, with such crystals developing alongside fruit ripening (Harris and Spurr 1969). Crystalline lycopene is the main form of the carotenoid found in health supplements (e.g., Johnson et al. 1997). This is important as the physical form of lycopene directly affects its subsequent antioxidant behaviour.

As a simple hydrocarbon, lycopene is highly hydrophobic and as a result orientates itself deep within the lipophilic core of lipid bilayers (Gruszecki 1999; Young and Lowe 2001). This contrasts with the behavior of more polar dietary carotenoids such as lutein and zeaxanthin, in which the functional oxygen-groups of these molecules serve to anchor them on opposing sides of a lipid membrane and thereby rigidify the membrane (Gruszecki 1999). This also raises interesting issues concerning how

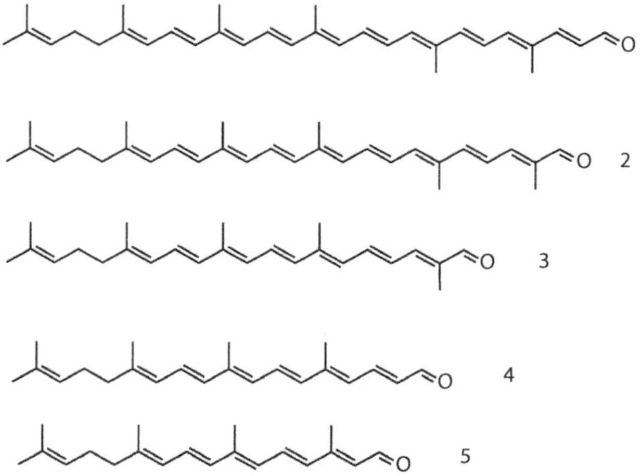

FIGURE 1.2 Structures of the common lycopenals found in tomatoes, tomato-based food-stuffs, and human plasma. (1) Apo-6′-lycopenal; (2) apo-8′-lycopenal; (3) apo-10′-lycopenal; (4) apo-14′-lycopenal; (5) apo-15′-lycopenal.

lycopene might interact with other antioxidants both in the lipid (e.g., other carot-enoids, vitamin E) and aqueous phases (e.g., vitamin C).

In addition to lycopene and its geometric isomers, several cleavage products (apo-lycopenoids) have been detected in raw tomatoes, tomato-derived foodstuffs, and most recently in human plasma. Apo-6′-lycopenal and apo-8′-lycopenal have been reported in raw tomatoes, and these and others (namely apo-10′-, apo-12′-, and apo-14′-lycopenals) have also been detected in human plasma (Kopec et al. 2010); see Figure 1.2 for structures. Ozonolysis of lycopene has been used to generate a range of apo-lycopenals (Kopec et al. 2010). The attribution of the beneficial effects of lycopene in the prevention of some diseases may in part be due to the properties of such compounds, whether naturally present in foods or generated as a result of metabolism (Catalano et al. 2013); see below.

1.2 ISOLATION AND ANALYSIS

Acyclic carotenes such as lycopene can readily aggregate in polar media, resulting in a potential reduction in biological activity. This is especially prevalent at high concentrations of lycopene. In cases where this might be an issue, the following can be done: the absorption spectrum of the solution can be checked for the presence of a prominent peak in the UV part of the spectrum (not to be confused by the so-called *cis*- or Z-peak—see Figure 1.3 for an example); the sample can be filtered to remove any micro-crystals of lycopene; or a few drops of dichloromethane can be added to disperse the crystals. General techniques for the isolation and handling of carotenoids including lycopene are given in Britton, Liaaen-Jensen and Pfander (1995). The UV/Vis spectroscopic characteristics of lycopene are given in Table 1.1.

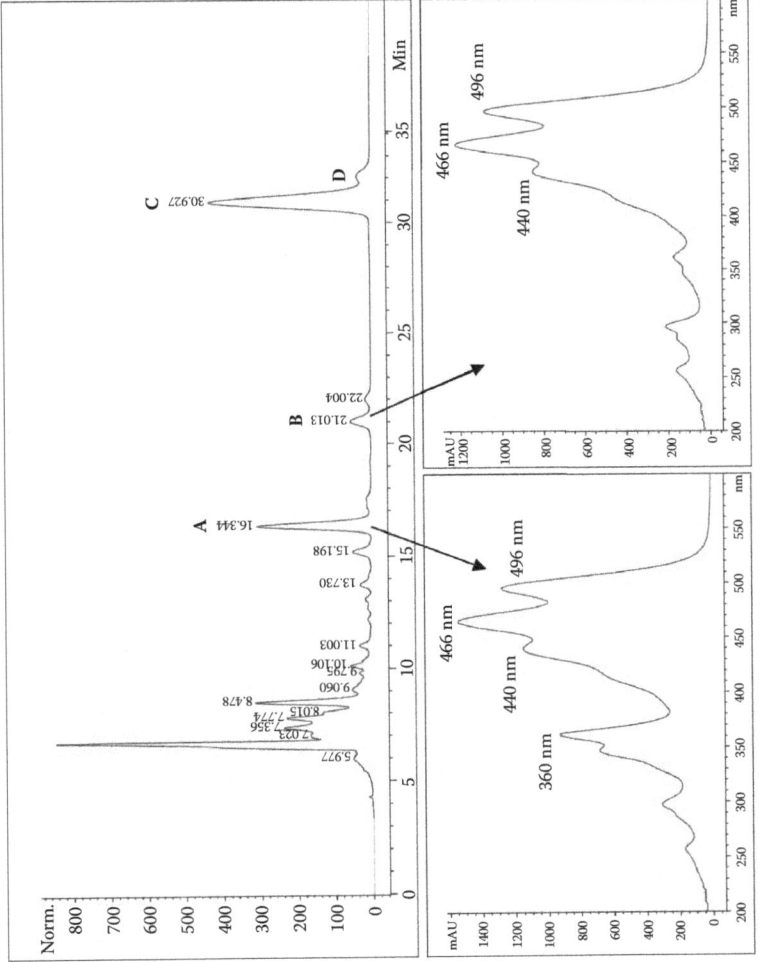

FIGURE 1.3 Separation of lycopene isomers by reversed-phase high-performance liquid chromatography using a C30 column. Full details of the chromatographic conditions are given in Graham et al. (2010). On-line absorption spectra are shown for (A) 13Z-lycopene and (B) 9Z-lycopene. Note that the 13Z-isomer displays a very prominent peak in the UV-region (λ_{max} 360 nm). The ratio of this peak to the main absorption maximum is semi-diagnostic (see Table 1.1). Other isomers identified are: (C) all-E-lycopene, and (D) 5Z-lycopene.

Additional spectroscopic characteristics (MS and NMR) of lycopene (Britton, Liaaen-Jensen, and Pfander 2004) have been published.

The separation and isolation of lycopene and its geometric isomers by high-performance liquid chromatography (HPLC) is a routine procedure that typically utilizes a C30 reversed-phase column as the stationary phase. For examples of chromatographic procedures routinely used for lycopene, its geometric isomers, cleavage, and oxidative products, see Graham et al. (2010) and Müller et al. (1997, 2011). Figure 1.3 shows the chromatographic separation of the most commonly observed geometric isomers of lycopene. The separation and detection of lycopenoids by HPLC-mass spectrometry has been described by Kopec et al. (2010) and Goupy et al. (2012). Spectroscopic characterization of some lycopenoids is provided by Goupy et al. (2012).

1.3   REACTIONS WITH REACTIVE OXYGEN SPECIES

The antioxidant properties of carotenoids are associated with their ability to scavenge or quench a range of reactive oxygen species, a topic reviewed by several authors (Bohm, Edge, and Truscott 2012; El-Agamey et al. 2004). The ability of carotenoids to quench singlet oxygen is arguably the best-known property (Di Mascio, Kaiser, and Sies 1989; Conn, Schalch, and Truscott 1991), but compounds such as lycopene interact with a wide range of radicals, including peroxyl radicals. The properties of the carotenoid radicals formed from such reactions are important, but, despite considerable recent interest in lycopene as a dietary antioxidant, to date only a few studies have been published concerning these reactions. Because of this, some of this section considers the general behavior of carotenoids, rather than specifically dealing with lycopene.

Di Mascio et al. (1989) were the first to demonstrate that lycopene possessed the highest singlet oxygen quenching rate of a range of carotenoids (carotenes and xanthophylls) and other antioxidants found in human plasma or dietary sources. Soon afterwards, Conn et al. (1991) also reported that lycopene had the highest rate constant for the quenching of singlet oxygen from a range of C40 carotenoids, but the differences between lycopene and β-carotene were not as great as previously reported. The reaction between carotenoids and singlet oxygen is reliant on the triplet energy level of the carotenoid, a function of the extent of the conjugated double bond system (11 C=C, in the case of lycopene). In an acyclic molecule such as lycopene, all 11 of these double bonds may be involved in reactions, but for carotenoids such as β-carotene (also 11 C=C) the cyclisation of the end-groups may reduce the efficiency due a loss of planarity of the two terminal double bonds (and a consequent increase in triplet energy level; Böhm et al. 2012). As mentioned above, acyclic carotenes such as lycopene have a strong tendency to aggregate (in polar environments and/or at high concentrations) and where this occurs the overall efficiency of singlet oxygen quenching would be substantially reduced.

Chain-breaking antioxidants such as carotenoids disrupt the chain reaction of lipid auto-oxidation by their ability to quench lipid peroxyl radicals. Lycopene and other carotenoids may scavenge peroxyl radicals (ROO•) in an initial step that involves one or more of: (a) electron transfer; (b) hydrogen abstraction and, (c) addition (see Equations 1.1–3).

Electron transfer $Car + ROO^{\bullet} \rightarrow Car^{\bullet +} + ROO^{-}$ (1.1)

Hydrogen abstraction $Car + ROO^{\bullet} \rightarrow Car^{\bullet} + ROOH$ (1.2)

Radical addition $Car + ROO^{\bullet} \rightarrow Car\text{-}ROO^{\bullet}$ (1.3)

Burton and Ingold (1984) first proposed that β-carotene scavenges peroxyl radicals by addition (Equation 1.3; yielding a resonance-stabilised carbon-centred peroxyl-carotenoid radical Car-ROO•), rather than by hydrogen abstraction (Equation 1.2). Analysis of the products formed as a result of the auto-oxidation of β–carotene suggested that scavenging of peroxyl radicals may occur both by these mechanisms as intermediates such as Car-ROO• (Kennedy and Liebler 1991; Liebler and McClure 1996; Mordi et al. 1991, 1993) and CAR• have been detected (Liebler and McClure 1996; Baker et al. 1999). Hydrogen abstraction leads to the formation of a carotenoid radical CAR• and a hydroperoxyl compound (Krinsky and Yeum 2003). ROO• can in fact interact right across the extensive polyene chain of a carotenoid such as lycopene, resulting in the formation of Car-ROO•. Woodall, Britton, and Jackson (1997) have suggested that all three of these mechanisms are possible. Electron transfer (Equation 1.1) is unlikely to be thermodynamically feasible for carotenoids such as lycopene that are embedded within a non-polar environment deep within a membrane, as charge separation will not be supported (El-Agamey et al. 2004). This is an example where observed behaviour *in vitro* and *in vivo* may differ. Electron transfer is more likely for xanthophylls where the polar substituents (typically –OH, in the case of dietary carotenoids) on one or both end groups of the carotenoid molecule may have the opportunity to interact with radicals in the aqueous phase (Woodall et al. 1997). This would explain why simple hydrocarbons such as β-carotene are less effective at preventing lipid peroxidation induced by a peroxyl radical initiator in the aqueous phase of a liposomal environment, whereas xanthophylls were able to afford protection against peroxyl radicals in both the aqueous and lipophilic regions.

The properties of the carotenoid radicals formed from the reactions described above determine the effectiveness of lycopene to function as an antioxidant and determine whether it could display pro-oxidant behaviour (Martin et al. 1999; Palozza 2004; Boehm et al. 2016). Similarly, the potentially harmful or beneficial effects of carotenoid products, which include a series of cleavage products (mainly apo-lycopenals and apo-lycopenols), oxidation products (primarily epoxides), and a number of geometric Z-isomers, will greatly influence its overall behaviour (Lowe, Vlismas, and Young 2003; Krinsky and Yeum 2003; Young, Phillip, and Lowe 2004; Graham et al. 2012). Figure 1.4 provides some of the oxidative intermediates of lycopene detected by Graham et al. (2012) and illustrates the possible pathways for lycopene when exposed to an oxidative scenario. Combinations of isomerisation (favouring

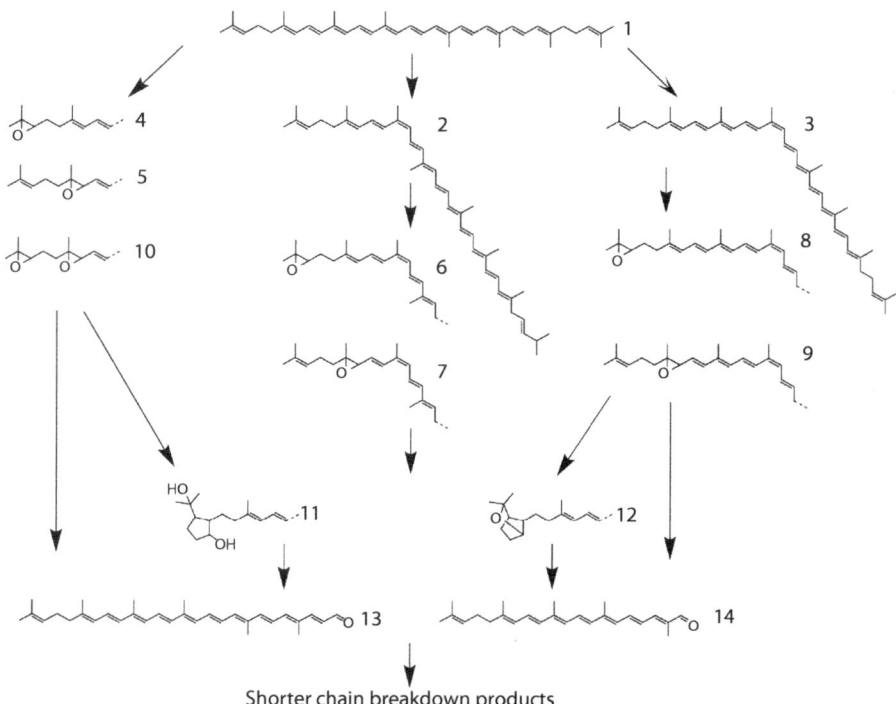

FIGURE 1.4 Possible pathways when all-*E*-lycopene is exposed to oxidative conditions such as cigarette smoke or Sin-1 (S-morpholinosydonimine) after Graham et al. (2012). (1) All-E lycopene; (2) 9Z-lycopene; (3) 13Z-lycopene; (4) lycopene-1,2-epoxide; (5) lycopene-5,6-epoxide; (6) 9Z-lycopene-1,2-epoxide; (7) 9Z-lycopene-5,6-epoxide; (8) 13Z-lycopene-1,2-epoxide; (9) 13Z-lycopene-5,6-epoxide; (10) 1,2,5,6-lycopene-di-epoxide; (11) 2,6-cyclolycopene-1,5-diol; (12) 2,6-cyclolycopene-1,5-epoxide; (13) apo-6′-lycopenal; (14) apo-12′-lycopenal.

thermodynamically stable forms), epoxidation, and cleavage of the polyene chain are all possible outcomes. The properties of many such products in biological systems remain poorly understood.

Yaping et al. (2002) reported a rate constant for the reaction of lycopene with the trichloromethylperoxyl radical $\left(CCl_3O_2^\bullet\right)$ of 1.17×10^9 M^{-1} s^{-1} (note that El-Agamey et al. (2004) believe that this value should be one order of magnitude lower). The published rate suggests a rapid reaction between the carotenoid molecule and the radical but El-Agamey et al. (2004) suggest that the very high concentrations of lycopene used in this study (in a polar environment) may result in its crystallization, compromising the resulting kinetics.

$$Lyc + Car^\bullet \rightarrow Lyc^\bullet + Car \qquad (1.4)$$

The carotenoid radical cations produced by these reactions are the most studied (see Böhm et al. 2012 for a review). The lycopene radical cation is particularly interesting as kinetic studies show that it is capable of reducing the radical cations of several other carotenoids, including β-carotene, canthaxanthin, zeaxanthin, and astaxanthin. This suggests that in a mix of such carotenoids (as would be normal in biological systems), lycopene would ultimately be sacrificed through the process of converting the other pigments back to the parent molecule (Equation 1.4). This would, however, require that these molecules are extremely closely aligned (e.g., within a membrane) and it is not always clear that this may occur because of the differences in structures and polarity (see above). Such a regeneration mechanism may serve to protect other carotenoids found in human plasma and tissues and thus indirectly afford protection from oxidation. This would fit well with the epidemiological observation that lycopene protects the human macula even though this particular carotenoid is not found in the eye (Mares-Perlman et al. 1995; Gouranton et al. 2008). Böhm and colleagues (2012) suggest that, should lutein and zeaxanthin (the main carotenoids of the macula) become oxidised, the resulting radical cations can then be re-converted back to the parent carotenoid by dietary lycopene.

Based on *in silico* modeling studies, Chasse et al. (2001) suggested that the peroxyl radical scavenging ability of the geometric isomers of lycopene was in the order 5Z > 9Z > 13Z > all-*E*. More recent *in vitro* work (Muller et al. 2011) partially confirmed this observation, but the results were often non-significant regarding the performance of different isomers. They further demonstrated that the radical scavenging activity of different geometric forms of lycopene was higher than that of both β-carotene and of α-tocopherol (up to several-fold). But no significant differences were found between different lycopene Z-isomers in terms of their ability to scavenge radicals via electron transfer reactions. The important implication from such studies is that the antioxidant ability of lycopene is not detrimentally affected by isomerisation (either as a result of food preparation or subsequent digestive processes), and may indeed be enhanced. Care is needed to ensure that any observed differences *in vitro* are not simply due to the improved solubility of Z-isomers compared to the all-*E* form of lycopene. This is particularly relevant due to the preferential accumulation of Z-isomers of lycopene in human plasma and tissues (see below).

1.4 REACTIONS WITH CIGARETTE SMOKE

Cigarette smoke contains over 80 known mutagenic carcinogenic agents, including arsenic, ammonia, formaldehyde, and benzopyrene (Nishikawa et al. 2004). The gaseous phase of cigarette smoke also contains free radicals along with both reactive oxygen substances and reactive nitrogenous substances, such as hydrogen peroxide, superoxide and hydroxyl radicals, and nitric oxide (van der Vliet and Cross 2000). Nitric oxide can react with superoxide to form peroxynitrite, a potent oxidising and nitrating compound that has been implicated in the initiation and progression of a number of chronic diseases including atherosclerosis (Beckman 1996; Dedon and Tannenbaum 2004). A wide range of reaction products of β-carotene (Lowe et al. 2009) and lycopene (Graham et al. 2012) are formed as a result of interaction with these components of cigarette smoke (Figure 1.4, Table 1.2). The 13Z- and 9Z-isomers of lycopene are major intermediates in the oxidative degradation of lycopene by cigarette smoke or 3-Morpholino-sydnonimine (Sin-1). A number of epoxides of lycopene are formed due to the interaction with the conjugated double bond chain. Graham et al. (2010) demonstrated that in isolated human plasma, levels of all-*E*-lycopene depleted to a greater extent than 5Z-lycopene or β-carotene, while in isolated LDL, both all-*E*- and 5Z-lycopene were shown to be more susceptible than β-carotene. Such results might support the sacrificial regenerative properties of lycopene in protecting other antioxidants.

Several products of lycopene have been formed by *in vitro* chemical oxidation (Kim et al. 2001; Caris-Veyrat et al. 2003; dos Anjos Ferreira et al. 2004; Rodriguez and Rodriguez-Amaya 2009). The properties of some of these are described below. In addition, several apo-lycopenals (see Figure 1.2 for structures) have been isolated from fruits, vegetables, and human plasma (Kopec et al. 2010). It has been shown that lycopene can be converted into apo-10′-lycopenal by intestinal carotenoid monoxygenase (CMO2) both *in vitro* and *in vivo* (Kiefer et al. 2001; Hu et al. 2006) and some of the products detected may arise from cleavage of 5Z- and 13Z-isomers of lycopene (Goupy et al. 2012).

Apo-10′-lycopenol has been detected in the lungs of ferrets (Hu et al. 2006) and apo-8′-lycopenal and apo-12′-lycopenal in the liver of rats (Gajic et al. 2006) after feeding with lycopene-enriched diets. A wide range of apolycopendoids have been detected in the plasma of humans following consumption of tomato juice: apo-6′-lycopenal, apo-8′-lycopenal, apo-10′-lycopenal, apo-12′-lycopenal, apo-14′-lycopenal, and apo-15′-lycopenal (Kopec et al. 2010).

TABLE 1.2
Oxidative Intermediates of Lycopene Produced by Exposure to Cigarette Smoke

T_R (min)	UV/Visible Spectral Maxima (nm)				III/II	Ab/All	Compound
8.5		433	458	489	0.23		2,6-cyclolycopene-1,5-diol
13.5	362	434	459	489	0.42	0.35	(Z)-2,6-cyclolycopene-1,5-epoxide
15.2	360	440	464	486		0.53	(13Z)-lycopene-1,2-epoxide
16.4	362	440	466	496	0.52	0.60	(13Z)-lycopene
17.2	360	436	460	489	0.33	0.36	lycopene-5,6-epoxide
18.0	362	434	459	488	0.23	0.20	unknown
20.7	362	440	466	498	0.68	0.16	(9Z)-lycopene
21.9	362	445	470	504	0.70	0.12	lycopene-1,2-epoxide
30.4	362	446	472	503	0.71	0.08	(all-*E*)-lycopene
31.6	362	446	472	503	0.73	0.09	(5Z)-lycopene

Note: Note that only all-*E* (98%) and 5Z-lycopene (1.5%) were detected in the starting material (Graham et al. 2012). See Figure 1.4 for structures.

1.5 PRO-OXIDANT PROPERTIES

It has been proposed that β-carotene (and, by implication, other dietary carotenoids such as lycopene) may act as a pro-oxidant under conditions of high concentrations of carotenoid at high partial pressures of oxygen (Burton and Ingold 1984). While β-carotene functions as an effective chain-breaking antioxidant, as the oxygen pressure is increased its efficacy declines, possibly due to auto-oxidative processes. In addition to a reduction in antioxidant effectiveness, carotenoids may also display a pro-oxidant effect, especially at high concentrations. Such behavior has been observed in *in vitro* cell studies (Palozza 1998; Lowe, Booth et al. 1999; Woods, Bilton, and Young 1999; Eichler, Sies, and Stahl 2002). More recently, the role of oxygen concentration on the ability of lycopene to protect lymphoid cells from γ-radiation has been explored by Boehm et al. (2016). A substantial reduction in the protective effect of this carotenoid was seen as the oxygen concentration was increased with a 50-fold difference in protection between studies performed at 0 or 100% oxygen. This further suggests that the antioxidant behavior of lycopene may be different in different biological tissues where the partial pressure of oxygen differs.

Work by Truscott and colleagues over the last 20 years has shown that the interaction between an oxidizing radical (e.g., $NO_2•$ as a component of cigarette smoke) and a carotenoid yields a carotenoid radical cation that is in itself a strong oxidizing species with a long lifetime (Burke et al. 2001). This may in turn lead to pro-oxidant effects in biological systems rather than the predicted antioxidant role. Such redox reactions need to be explored more fully, especially using combinations of dietary carotenoids and other dietary antioxidants such as vitamin C. To minimize oxidative damage from such species it may be that a high concentration of vitamin C is needed to regenerate the carotenoid radical cation and minimize the resulting oxidative damage (Edge and Truscott 1999).

1.6 LYCOPENE INTAKE

Many studies in the 1990s and early 2000s associated plasma lycopene concentrations with the intake of fruit and vegetables (McEligot et al. 1999; van Kappel, Martinez-Garcia et al. 2001; van Kappel, Steghens et al. 2001). However, in the Western world the development of convenience food and fast food has resulted in more processed tomatoes being consumed, along with lycopene. However, in more recent studies the uptake of lycopene in humans was not attributed to fruit or vegetable intake (Zhou et al. 2016; Couillard et al. 2016). In the study by Zhou it was determined that in the U.S.A. the majority of lycopene in the diet is obtained through pizza and pasta rather than fruit and vegetables. The concentration of lycopene in processed foods is far higher than in fruit or vegetables. A study by Singh and Goyal (2008) identified that pizza sauce contained lycopene at 32.9 mg/g wet weight, which was the equivalent of 9.9 mg per 125 g serving. For watermelon it was 4.1 mg per 100 g wet weight, and 11.5 mg per 280 g serving.

In a ripening tomato the conversion of the chloroplast to the chromoplast is associated with the accumulation of carotenoids, volatiles, and organic acids (Klee and Giovannoni 2011). Lycopene can form crystalline structures or may have more complex interactions with proteins within the chloroplasts during tomato ripening (Parada and Aguilera 2007). This complex arrangement leads to poor bioavailability of lycopene in raw tomatoes compared to processed tomatoes (Gartner, Stahl, and Sies 1997). Inappropriate storage and handling of tomatoes can also affect the concentration, flavor, and isomer profiles of lycopene within the chromoplasts. Exposure to oxygen or light may bleach lycopene or promote isomerization (Xianquan et al. 2005). The modern processing of tomatoes has been assessed and thermal processing leads to an improved bioavailability of lycopene, without significant isomerisation. The impact of heat affects the integrity of the chromoplasts and allows release from the tissue matrix (Colle et al. 2010).

The amount of lycopene absorbed following a meal seems to be constrained by its incorporation into micelles in the duodenum (Tyssandier et al. 2003). Another study indicated that the amount absorbed was irrespective of the dose of lycopene (Diwadkar-Navsariwala et al. 2003). Many studies have attempted to assess the nutritional intake of lycopene in various populations (Table 1.3).

When consuming products rich in lycopene, the bioavailability of the carotenoid in the gut is dependent on two important factors. First, how the tomato products were processed (especially heating). Second, how much fat is consumed alongside the lycopene. This was demonstrated in one study (Brown et al. 2004) where salads rich in carotenoids were consumed with full-fat dressings compared to those with a reduced-fat dressing. In the case of the full-fat dressing, the volunteers had greater plasma concentrations of carotenoids. In the stomach, the lycopene is released through the action of pepsin on the chromoplasts, and it was thought (Re et al. 2001) that the acidic pH of the stomach may influence the degree of isomerization of lycopene. However, various *in vitro* models suggest that no isomerization occurs in the stomach (Richelle et al. 2010). The situation is made even more complex as isomerization is not permanent and Z-isomers—particularly the $13Z$ isomer—are thermodynamically unstable and can revert back to the all-E state (Moraru and Lee 2005). When digestion is complete, the chyme is released into the duodenum, the pH increases and a complex interaction between carotenoids, fatty acids, and bile acids occurs. This results in the formation of micelles, incorporating lycopene. Micelles are composed of a shell that is hydrophilic in nature, whilst the hydrophobic core is where lycopene

TABLE 1.3
Daily Dietary Intakes of Lycopene Across the World

Country	Mean Intake mg/day
USA	Men 6.6–10.5; women 5.7–10.4
Italy	7.4
Spain	1.6
UK	1.1
Australia	3.8

Source: Data obtained from Porrini, M., and P. Riso. 2005. "What are typical lycopene intakes?" *J Nutr* 135 (8):2042s–5s.

would be found. However, the structure of Z-isomers allows for their preferential incorporation into micelles over all-E lycopene (Failla, Chitchumroonchokchai, and Ishida 2008; Boileau, Boileau, and Erdman 2002). Although the majority of lycopene will be in the all-E form, this process will be selective for the Z-isomers.

The micelles then interact with transporters present on the surface of the enterocytes in the gut. Several candidate transporters have been identified (During, Dawson, and Harrison 2005), including Scavenger Receptor class B-type I (SR-BI), Niemann Pick C1-like 1 (NPC1L1) and ATP-binding cassette transporter (ABCA1). Further work (Moussa et al. 2008) using a Caco-2 monolayer and a specific human anti-SR-BI antibody reduced the uptake of all-E and 5Z-lycopene in the *in vitro* study, whereas a similar approach for NPC1L1 had no effect. In transgenic mice that overexpressed SR-BI in the enterocytes the plasma concentrations of lycopene were greater that the control groups. This study led to the suggestion that SR-BI was the more important transporter for the uptake of lycopene to the enterocytes.

In 2007 an association between the serum concentration of lycopene and SNPs was established (Borel et al. 2015, Borel et al. 2009, Borel et al. 2007). This led to the supposition that the ability for an individual to respond to lycopene may be influenced by 28 SNPs in 16 genes. This may partially explain the variability in lycopene serum concentrations amongst individuals taking lycopene supplements, or consuming a lycopene-rich diet. The uptake of lycopene is determined by the fat content of the meal and genetic determinants, and may also be affected by the microflora of the gut, as a probiotic yoghurt reduced plasma concentrations in 17 women over a four-week period (Fabian and Elmadfa 2007).

1.7 TRANSPORT

Once the micelles from the gut containing lycopene are absorbed into the enterocyte they are processed to form chylomicrons, which enter the lymphatic system and eventually enter the blood stream. The chylomicrons are processed through the action of lipoprotein lipase and the remnants are processed in the liver to form Very-Low-Density Lipoproteins (VLDL). These lipoproteins are processed to form Intermediate-Density Lipoproteins (IDL) and Low-Density Lipoproteins (LDL) following the actions of lipoprotein lipase and the enrichment of the lipoproteins with cholesterol. Studies have indicated that lycopene and other carotenoids are present in chylomicrons, VLDL, and LDL (Maruyama et al. 2001; Lin et al. 2000; Lowe, Bilton et al. 1999). The LDL particle is composed of apolipoprotein B100, lipids, and antioxidants, including α-tocopherol, β-carotene, and lycopene. Two independent studies indicated that there are approximately 14 molecules of α-tocopherol, one molecule of lycopene, and only 0.5 molecules of β-carotene per LDL particle (Milde, Elstner, and Grassmann 2007; Romanchik, Morel, and Harrison 1995). The expression of LDL receptors on tissues allows the uptake of lycopene incorporated within LDL particles into cells. Human organs store lycopene to varying degrees. Lycopene is found in the highest concentrations in the liver, testes, adrenal glands, and adipose tissues (Kun, Ssonko Lule, and Xiao-Lin 2006). It is found in lower concentrations in the kidney, ovary, lung, and prostate (Kun, Ssonko Lule, and Xiao-Lin 2006).

1.8 METABOLISM

In 2001, using cDNA technology, it was shown that humans can express CMO I (carotene-15,15′-monooxygenase) and CMO II (carotene-9,10-monooxygenase) (Kiefer et al. 2001). The same study also indicated that these enzymes were expressed in various tissues in the mouse, including the small intestine, brain, testes, and liver. CMO I has the ability to cleave β-carotene centrally to provide two molecules of retinal, while CMO II cleaves non-provitamin A carotenoids such as lycopene asymmetrically to yield apo-8′-lycopenal and apo-12′-lycopenal (Ford et al. 2010).

In vitro and *in vivo* studies indicate that lycopene and/or its metabolites may act via nuclear receptors and transcription systems including PPARs, liver X receptor, and nuclear factor erythroid-derived 2 (NRF2) to induce downstream physiologic changes (Zaripheh et al. 2006; Yang et al. 2012; Lian and Wang 2008; Ben-Dor et al. 2005). In particular, apo-8′-lycopenal acts via NRF2 to affect phase-II enzyme expression in the liver (Chung et al. 2012). However, it may be that some of these apo-lycopenals may also be derived from the diet (Kopec et al. 2010). A mixture of lycopenals, namely, apo-6′-, apo-8′-, apo-10′-, apo-12′-, and apo-14′-lycopenals, were found in processed and raw foods. The concentration of lycopenals in plasma is about 0.2% of the total plasma concentration of lycopenes at about 2 nM (Kopec et al. 2010).

1.9 CONCLUSIONS

For many years, the focus on the potential health benefits of carotenoids has focused on β-carotene. More recently, such research has encompassed the acyclic compound lycopene, a major constituent of the human diet via its ingestion from tomatoes and tomato-based products. The structure of lycopene with its conjugated backbone means that, under appropriate conditions, it can be a highly effective antioxidant. At the same time, however, its structure means that it has a strong tendency to crystallize, which can significantly reduce its potential effectiveness. The extended double bond chain that forms the chromophore of lycopene means that it is prone to processes such as isomerization during food preparation, digestion, or as a result of oxidative attack. The properties of these geometric isomers and other oxidative byproducts (e.g., epoxides or cleavage compounds such as lycopenals) are, however, by comparison with the native compound, poorly understood and worthy of additional research efforts.

ACKNOWLEDGMENTS

We thank George Britton and George Truscott for their advice and support on aspects of carotenoid research over many years. GL and DG thank Boots UK and the European Commission (Lycocard 016213) for supporting their recent research on lycopene.

REFERENCES

Aebischer, C. P., J. Schierle, and W. Schuep. 1999. "Simultaneous determination of retinol, tocopherols, carotene, lycopene and xanthophylls in plasma by means of reversed-phase high-performance liquid chromatography." *Methods Enzymol* 299:348–62.

Baker, D. L., E. S. Krol, N. Jacobsen, and D. C. Liebler. 1999. "Reactions of beta-carotene with cigarette smoke oxidants. Identification of carotenoid oxidation products and evaluation of the prooxidant/antioxidant effect." *Chem Res Toxicol* 12 (6):535–43. doi: 10.1021/tx980263v.

Beckman, J. S. 1996. "Oxidative damage and tyrosine nitration from peroxynitrite." *Chem Res Toxicol* 9 (5):836–44. doi: 10.1021/tx9501445.

Ben-Dor, A., M. Steiner, L. Gheber, M. Danilenko, N. Dubi, K. Linnewiel, A. Zick, Y. Sharoni, and J. Levy. 2005. "Carotenoids activate the antioxidant response element transcription system." *Mol Cancer Ther* 4 (1):177–86.

Berg, M. A., G. A. Chasse, E. Deretey, A. K. Füzéry, B. M. Fung, D. Y. K. Fung, H. Henry-Riyad, A. C. Lin, M. L. Mak, A. Mantas, M. Patel, I. V. Repyakh, M. Staikova, S. J. Salpietro, Ting-Hua Tang, J. C. Vank, A. Perczel, G. I. Csonka, Ö Farkas, L. L. Torday, Z. Székely, and I. G. Csizmadia. 2000. "Prospects in computational molecular medicine: A millennial mega-project on peptide folding." *Journal of Molecular Structure: THEOCHEM* 500 (1–3):5–58. doi: 10.1016/S0166-1280(00)00448-6.

Boehm, F., R. Edge, T. G. Truscott, and C. Witt. 2016. "A dramatic effect of oxygen on protection of human cells against gamma-radiation by lycopene." *FEBS Lett* 590 (8):1086–93. doi: 10.1002/1873-3468.12134.

Bohm, F., R. Edge, and T. G. Truscott. 2012. "Interactions of dietary carotenoids with singlet oxygen ($1O2$) and free radicals: Potential effects for human health." *Acta Biochim Pol* 59 (1):27–30.

Boileau, T. W., A. C. Boileau, and J. W. Erdman, Jr. 2002. "Bioavailability of all-trans and cis-isomers of lycopene." *Exp Biol Med (Maywood)* 227 (10):914–19.

Borel, P., C. Desmarchelier, M. Nowicki, and R. Bott. 2015. "Lycopene bioavailability is associated with a combination of genetic variants." *Free Radic Biol Med* 83:238–44. doi: 10.1016/j.freeradbiomed.2015.02.033.

Borel, P., M. Moussa, E. Reboul, B. Lyan, C. Defoort, S. Vincent-Baudry, M. Maillot, M. Gastaldi, M. Darmon, H. Portugal, D. Lairon, and R. Planells. 2009. "Human fasting plasma concentrations of vitamin E and carotenoids, and their association with genetic variants in apo C-III, cholesteryl ester transfer protein, hepatic lipase, intestinal fatty acid binding protein and microsomal triacylglycerol transfer protein." *Br J Nutr* 101 (5):680–7. doi: 10.1017/s0007114508030754.

Borel, P., M. Moussa, E. Reboul, B. Lyan, C. Defoort, S. Vincent-Baudry, M. Maillot, M. Gastaldi, M. Darmon, H. Portugal, R. Planells, and D. Lairon. 2007. "Human plasma levels of vitamin E and carotenoids are associated with genetic polymorphisms in genes involved in lipid metabolism." *J Nutr* 137 (12):2653–9.

Britton, G., S. Liaaen-Jensen, and H. Pfander. 1995. *Carotenoids. Vol 1B Spectroscopy.* Basel: Birkhäuser Verlag.

Britton, G., S. Liaaen-Jensen, and H. Pfander. 2004. *Carotenoids Handbook.* Basel: Birkhäuser Verlag.

Brown, M. J., M. G. Ferruzzi, M. L. Nguyen, D. A. Cooper, A. L. Eldridge, S. J. Schwartz, and W. S. White. 2004. "Carotenoid bioavailability is higher from salads ingested with full-fat than with fat-reduced salad dressings as measured with electrochemical detection." *Am J Clin Nutr* 80 (2):396–403.

Burke, M., R. Edge, E. J. Land, and T. G. Truscott. 2001. "Characterisation of carotenoid radical cations in liposomal environments: Interaction with vitamin C." *J Photochem Photobiol B* 60 (1):1–6.

Burton, G. W., and K. U. Ingold. 1984. "Beta-Carotene: An unusual type of lipid antioxidant." *Science* 224 (4649):569–73. doi: 10.1126/science.6710156.

Caris-Veyrat, C., A. Schmid, M. Carail, and V. Bohm. 2003. "Cleavage products of lycopene produced by in vitro oxidations: Characterization and mechanisms of formation." *J Agric Food Chem* 51 (25):7318–25. doi: 10.1021/jf034735+.

Catalano, A., R. E. Simone, A. Cittadini, E. Reynaud, C. Caris-Veyrat, and P. Palozza. 2013. "Comparative antioxidant effects of lycopene, apo-10'-lycopenoic acid and apo-14'-lycopenoic acid in human macrophages exposed to H2O2 and cigarette smoke extract." *Food Chem Toxicol* 51:71–9. doi: 10.1016/j.fct.2012.08.050.

Chasse, G. A., K. P. Chasse, A. Kucsman, L. L. Torday, and J. G. Papp. 2001. "Conformational potential energy surfaces of a Lycopene model." *Journal of Molecular Structure: THEOCHEM* 571 (1–3):7–26. doi: http://dx.doi.org/10.1016/S0166-1280(01)00413-4.

Chung, J., K. Koo, F. Lian, K. Q. Hu, H. Ernst, and X. D. Wang. 2012. "Apo-10'-lycopenoic acid, a lycopene metabolite, increases sirtuin 1 mRNA and protein levels and decreases hepatic fat accumulation in ob/ob mice." *J Nutr* 142 (3):405–10. doi: 10.3945 /jn.111.150052.

Colle, I., L. Lemmens, S. Van Buggenhout, A. Van Loey, and M. Hendrickx. 2010. "Effect of thermal processing on the degradation, isomerization, and bioaccessibility of lycopene in tomato pulp." *J Food Sci* 75 (9):C753–9. doi: 10.1111/j.1750-3841.2010.01862.x.

Conn, P. F., W. Schalch, and T. G. Truscott. 1991. "The singlet oxygen and carotenoid interaction." *J Photochem Photobiol B* 11 (1):41–7.

Couillard, C., S. Lemieux, M. C. Vohl, P. Couture, and B. Lamarche. 2016. "Carotenoids as biomarkers of fruit and vegetable intake in men and women." *Br J Nutr* 116 (7):1206–15. doi: 10.1017/s0007114516003056.

Dedon, P. C., and S. R. Tannenbaum. 2004. "Reactive nitrogen species in the chemical biology of inflammation." *Arch Biochem Biophys* 423 (1):12–22.

Di Mascio, P., S. Kaiser, and H. Sies. 1989. "Lycopene as the most efficient biological carotenoid singlet oxygen quencher." *Arch Biochem Biophys* 274 (2):532–8.

Diwadkar-Navsariwala, V., J. A. Novotny, D. M. Gustin, J. A. Sosman, K. A. Rodvold, J. A. Crowell, M. Stacewicz-Sapuntzakis, and P. E. Bowen. 2003. "A physiological pharmacokinetic model describing the disposition of lycopene in healthy men." *J Lipid Res* 44 (10):1927–39. doi: 10.1194/jlr.M300130-JLR200.

dos Anjos Ferreira, A. L., K. J. Yeum, R. M. Russell, N. I. Krinsky, and G. Tang. 2004. "Enzymatic and oxidative metabolites of lycopene." *J Nutr Biochem* 15 (8):493–502.

During, A., H. D. Dawson, and E. H. Harrison. 2005. "Carotenoid transport is decreased and expression of the lipid transporters SR-BI, NPC1L1, and ABCA1 is downregulated in Caco-2 cells treated with ezetimibe." *J Nutr* 135 (10):2305–12.

Edge, R., and T. G. Truscott. 1999. "Carotenoid Radicals and the Interaction of Carotenoids with Active Oxygen Species." In *The Photochemistry of Carotenoids*, edited by H. A. Frank, A. J. Young, G. Britton, and R. J. Cogdell, 223–34. Dordrecht: Springer Netherlands.

Eichler, O., H. Sies, and W. Stahl. 2002. "Divergent optimum levels of lycopene, beta-carotene and lutein protecting against UVB irradiation in human fibroblastst." *Photochem Photobiol* 75 (5):503–506.

El-Agamey, A., G. M. Lowe, D. J. McGarvey, A. Mortensen, D. M. Phillip, T. G. Truscott, and A. J. Young. 2004. "Carotenoid radical chemistry and antioxidant/pro-oxidant properties." *Arch Biochem Biophys* 430 (1):37–48. doi: 10.1016/j.abb.2004.03.007.

Fabian, E., and I. Elmadfa. 2007. "The effect of daily consumption of probiotic and conventional yoghurt on oxidant and anti-oxidant parameters in plasma of young healthy women." *Int J Vitam Nutr Res* 77 (2):79–88. doi: 10.1024/0300-9831.77.2.79.

Failla, M. L., C. Chitchumroonchokchai, and B. K. Ishida. 2008. "In vitro micellarization and intestinal cell uptake of cis isomers of lycopene exceed those of all-trans lycopene." *J Nutr* 138 (3):482–6.

Ford, N. A., S. K. Clinton, J. von Lintig, A. Wyss, and J. W. Erdman, Jr. 2010. "Loss of carotene-9',10'-monooxygenase expression increases serum and tissue lycopene concentrations in lycopene-fed mice." *J Nutr* 140 (12):2134–8. doi: 10.3945/jn.110.128033.

Gajic, M., S. Zaripheh, F. Sun, and J. W. Erdman, Jr. 2006. "Apo-8'-lycopenal and apo-12'-lycopenal are metabolic products of lycopene in rat liver." *J Nutr* 136 (6):1552–7.

Gartner, C., W. Stahl, and H. Sies. 1997. "Lycopene is more bioavailable from tomato paste than from fresh tomatoes." *Am J Clin Nutr* 66 (1):116–22.

Goupy, P., E. Reynaud, O. Dangles, and C. Caris-Veyrat. 2012. "Antioxidant activity of (all-E)-lycopene and synthetic apo-lycopenoids in a chemical model of oxidative stress in the gastro-intestinal tract." *New Journal of Chemistry* 36 (3):575–87. doi: 10.1039/C1NJ20437H.

Gouranton, E., C. E. Yazidi, N. Cardinault, M. J. Amiot, P. Borel, and J. F. Landrier. 2008. "Purified low-density lipoprotein and bovine serum albumin efficiency to internalise lycopene into adipocytes." *Food Chem Toxicol* 46 (12):3832–6. doi: 10.1016/j.fct.2008.10.006.

Graham, D. L., M. Carail, C. Caris-Veyrat, and G. M. Lowe. 2010. "Cigarette smoke and human plasma lycopene depletion." *Food Chem Toxicol* 48:2413–2420. doi: 10.1016/j.fct.2010.06.001.

Graham, D. L., M. Carail, C. Caris-Veyrat, and G. M. Lowe. 2012. "(13Z)- and (9Z)-lycopene isomers are major intermediates in the oxidative degradation of lycopene by cigarette smoke and Sin-1." *Free Radic Res* 46 (7):891–902. doi: 10.3109/10715762.2012.686663.

Gruszecki, W. I. 1999. "Carotenoids in Membranes." In *The Photochemistry of Carotenoids*, edited by H. A. Frank, A. J. Young, G. Britton, and R. J. Cogdell, 363–79. Dordrecht: Springer Netherlands.

Harris, W. M., and A. R. Spurr. 1969. "Chromoplasts of Tomato Fruits. I. Ultrastructure of Low-Pigment and High-Beta Mutants. Carotene Analyses." *American Journal of Botany* 56 (4):369–79. doi: 10.2307/2440812.

Hengartner, U., K. Bernhard, K. Meyer, G. Englert, and E. Glinz. 1992. "Synthesis, Isolation, and NMR-Spectroscopic Characterization of Fourteen (Z)-Isomers of Lycopene and of Some Acetylenic Didehydro- and Tetradehydrolycopenes." *Helvetica Chimica Acta* 75 (6):1848–65. doi: 10.1002/hlca.19920750611.

Hu, K. Q., C. Liu, H. Ernst, N. I. Krinsky, R. M. Russell, and X. D. Wang. 2006. "The biochemical characterization of ferret carotene-9',10'-monooxygenase catalyzing cleavage of carotenoids in vitro and in vivo." *J Biol Chem* 281 (28):19327–38. doi: 10.1074/jbc.M512095200.

Jenab, M., P. Ferrari, M. Mazuir, A. Tjonneland, F. Clavel-Chapelon, J. Linseisen, A. Trichopoulou, R. Tumino, H. B. Bueno-de-Mesquita, E. Lund, C. A. Gonzalez, G. Johansson, T. J. Key, and E. Riboli. 2005. "Variations in lycopene blood levels and tomato consumption across European countries based on the European Prospective Investigation into Cancer and Nutrition (EPIC) study." *J Nutr* 135 (8):2032s–6s.

Johnson, E. J., J. Qin, N. I. Krinsky, R. M. Russell. 1997. "Ingestion by men of a combined dose of beta-carotene and lycopene does not affect the absorption of beta-carotene but improves that of lycopene." *J Nutr* 127(9):1833–7.

Kennedy, T. A., and D. C. Liebler. 1991. "Peroxyl radical oxidation of beta-carotene: Formation of beta-carotene epoxides." *Chem Res Toxicol* 4 (3):290–5.

Kiefer, C., S. Hessel, J. M. Lampert, K. Vogt, M. O. Lederer, D. E. Breithaupt, and J. von Lintig. 2001. "Identification and characterization of a mammalian enzyme catalyzing the asymmetric oxidative cleavage of provitamin A." *J Biol Chem* 276 (17):14110–16. doi: 10.1074/jbc.M011510200.

Kim, S. J., E. Nara, H. Kobayashi, J. Terao, and A. Nagao. 2001. "Formation of cleavage products by autoxidation of lycopene." *Lipids* 36 (2):191–9.

Klee, H. J., and J. J. Giovannoni. 2011. "Genetics and control of tomato fruit ripening and quality attributes." *Annu Rev Genet* 45:41–59. doi: 10.1146/annurev-genet-110410-132507.

Kopec, R. E., K. M. Riedl, E. H. Harrison, R. W. Curley, Jr., D. P. Hruszkewycz, S. K. Clinton, and S. J. Schwartz. 2010. "Identification and quantification of apo-lycopenals in fruits, vegetables, and human plasma." *J Agric Food Chem* 58 (6):3290–6. doi: 10.1021/jf100415z.

Krinsky, N. I., and K. J. Yeum. 2003. "Carotenoid-radical interactions." *Biochem Biophys Res Commun* 305 (3):754–60.

Kun, Y., U. S. Lule, and D. Xiao-Lin. 2006. "Lycopene: Its Properties and Relationship to Human Health." *Food Reviews International* 22 (4):309–33. doi: 10.1080/87559120600864753.

Lian, F., and X. D. Wang. 2008. "Enzymatic metabolites of lycopene induce Nrf2-mediated expression of phase II detoxifying/antioxidant enzymes in human bronchial epithelial cells." *Int J Cancer* 123 (6):1262–8. doi: 10.1002/ijc.23696.

Liebler, D. C., and T. D. McClure. 1996. "Antioxidant reactions of beta-carotene: Identification of carotenoid-radical adducts." *Chem Res Toxicol* 9 (1):8–11. doi: 10.1021/tx950151t.

Lin, S., L. Quaroni, W. S. White, T. Cotton, and G. Chumanov. 2000. "Localization of carotenoids in plasma low-density lipoproteins studied by surface-enhanced resonance Raman spectroscopy." *Biopolymers* 57 (4):249–56. doi: 10.1002/1097-0282(2000) 57:4<249::aid-bip6>3.0.co;2-1.

Lowe, G. M., R. F. Bilton, I. G. Davies, T. C. Ford, D. Billington, and A. J. Young. 1999. "Carotenoid composition and antioxidant potential in subfractions of human low-density lipoprotein." *Ann Clin Biochem* 36 (Pt 3):323–32.

Lowe, G. M., L. A. Booth, A. J. Young, and R. F. Bilton. 1999. "Lycopene and beta-carotene protect against oxidative damage in HT29 cells at low concentrations but rapidly lose this capacity at higher doses." *Free Radic Res* 30 (2):141–51.

Lowe, G. M., K. Vlismas, D. L. Graham, M. Carail, C. Caris-Veyrat, and A. J. Young. 2009. "The degradation of (all-E)-beta-carotene by cigarette smoke." *Free Radic Res* 43 (3):280–6. doi: 10.1080/10715760802691497.

Lowe, G. M., K. Vlismas, and A. J. Young. 2003. "Carotenoids as prooxidants?" *Mol Aspects Med* 24 (6):363–9.

Mares-Perlman, J. A., W. E. Brady, R. Klein, B. E. Klein, P. Bowen, M. Stacewicz-Sapuntzakis, and M. Palta. 1995. "Serum antioxidants and age-related macular degeneration in a population-based case-control study." *Arch Ophthalmol* 113 (12):1518–23.

Martin, H. D., C. Ruck, M. Schmidt, S. Sell, S. Beutner, B. Mayer, and R. Walsh. 1999. "Chemistry of carotenoid oxidation and free radical reactions." *Pure Appl Chem* 71: 2253–62. doi: 10.1351/pac199971122253.

Maruyama, C., K. Imamura, S. Oshima, M. Suzukawa, S. Egami, M. Tonomoto, N. Baba, M. Harada, M. Ayaori, T. Inakuma, and T. Ishikawa. 2001. "Effects of tomato juice consumption on plasma and lipoprotein carotenoid concentrations and the susceptibility of low density lipoprotein to oxidative modification." *J Nutr Sci Vitaminol (Tokyo)* 47 (3):213–21.

McEligot, A. J., C. L. Rock, S. W. Flatt, V. Newman, S. Faerber, and J. P. Pierce. 1999. "Plasma carotenoids are biomarkers of long-term high vegetable intake in women with breast cancer." *J Nutr* 129 (12):2258–63.

Milde, J., E. F. Elstner, and J. Grassmann. 2007. "Synergistic effects of phenolics and carotenoids on human low-density lipoprotein oxidation." *Mol Nutr Food Res* 51 (8):956–61. doi: 10.1002/mnfr.200600271.

Moraru, C., and T. C. Lee. 2005. "Kinetic studies of lycopene isomerization in a tributyrin model system at gastric pH." *J Agric Food Chem* 53 (23):8997–9004. doi: 10.1021 /jf051672h.

Mordi, R. C., J. C. Walton, G. W. Burton, L. Hughes, K. U. Ingold, and D. A. Lindsay. 1991. "Exploratory study of β-carotene autoxidation." *Tetrahedron Letters* 32 (33):4203–6.

Mordi, R. C., J. C. Walton, G. W. Burton, L. Hughes, K. U. Ingold, L. A. David, and D. J. Moffatt. 1993. "Oxidative degradation of β-carotene and β-apo-8'-carotenal." *Tetrahedron* 49 (4):911–28.

Moussa, M., J. F. Landrier, E. Reboul, O. Ghiringhelli, C. Comera, X. Collet, K. Frohlich, V. Bohm, and P. Borel. 2008. "Lycopene absorption in human intestinal cells and in mice involves scavenger receptor class B type I but not Niemann-Pick C1-like 1." *J Nutr* 138 (8):1432–6.

Muller, L., P. Goupy, K. Frohlich, O. Dangles, C. Caris-Veyrat, and V. Bohm. 2011. "Comparative study on antioxidant activity of lycopene (Z)-isomers in different assays." *J Agric Food Chem* 59 (9):4504–11. doi: 10.1021/jf1045969.

Muller, R. K., K. Bernhard, A. Giger, G. Moine, and U. Hengartner. 1997. "(E/Z)-Isomeric carotenes." *Pure Appl Chem* 69:2039–46.

Nishikawa, A., Y. Mori, I. S. Lee, T. Tanaka, and M. Hirose. 2004. "Cigarette smoking, metabolic activation and carcinogenesis." *Curr Drug Metab* 5 (5):363–73.

Palozza, P. 1998. "Prooxidant actions of carotenoids in biologic systems." *Nutr Rev* 56 (9):257–65.

Palozza, P. 2004. "Evidence for Pro-Oxidant Effects of Carotenoids *In Vitro* and *In Vivo.*" In *Carotenoids in Health and Disease*, 127–49. CRC Press.

Parada, J., and J. M. Aguilera. 2007. "Food microstructure affects the bioavailability of several nutrients." *J Food Sci* 72 (2):R21–32. doi: 10.1111/j.1750-3841.2007.00274.x.

Porrini, M., and P. Riso. 2005. "What are typical lycopene intakes?" *J Nutr* 135 (8):2042s–5s.

Re, R., P. D. Fraser, M. Long, P. M. Bramley, and C. Rice-Evans. 2001. "Isomerization of lycopene in the gastric milieu." *Biochem Biophys Res Commun* 281 (2):576–81. doi: 10.1006/bbrc.2001.4366.

Richelle, M., B. Sanchez, I. Tavazzi, P. Lambelet, K. Bortlik, and G. Williamson. 2010. "Lycopene isomerisation takes place within enterocytes during absorption in human subjects." *Br J Nutr* 103 (12):1800–7. doi: 10.1017/s0007114510000103.

Rodriguez, E. B., and D. B. Rodriguez-Amaya. 2009. "Lycopene epoxides and apo-lycopenals formed by chemical reactions and autoxidation in model systems and processed foods." *J Food Sci* 74 (9):C674–82. doi: 10.1111/j.1750-3841.2009.01353.x.

Romanchik, J. E., D. W. Morel, and E. H. Harrison. 1995. "Distributions of carotenoids and alpha-tocopherol among lipoproteins do not change when human plasma is incubated in vitro." *J Nutr* 125 (10):2610–17.

Schierle, J., W. Bretzel, I. Bühler, N. Faccin, D. Hess, K. Steiner, and W. Schüep. 1997. "Content and isomeric ratio of lycopene in food and human blood plasma." *Food Chemistry* 59 (3):459–65. doi:10.1016/S0308-8146(96)00177-X.

Singh, P., and G. K. Goyal. 2008. "Dietary Lycopene: Its Properties and Anticarcinogenic Effects." *Comprehensive Reviews in Food Science and Food Safety* 7 (3):255–70. doi: 10.1111/j.1541-4337.2008.00044.x.

Tyssandier, V., E. Reboul, J. F. Dumas, C. Bouteloup-Demange, M. Armand, J. Marcand, M. Sallas, and P. Borel. 2003. "Processing of vegetable-borne carotenoids in the human stomach and duodenum." *Am J Physiol Gastrointest Liver Physiol* 284 (6):G913–23. doi: 10.1152/ajpgi.00410.2002.

van der Vliet, A., and C. E. Cross. 2000. "Oxidants, nitrosants, and the lung." *Am J Med* 109 (5):398–421.

van Kappel, A. L., C. Martinez-Garcia, S. Elmstahl, J. P. Steghens, V. Chajes, F. Bianchini, R. Kaaks, and E. Riboli. 2001. "Plasma carotenoids in relation to food consumption in Granada (southern Spain) and Malmo (southern Sweden)." *Int J Vitam Nutr Res* 71 (2):97–102. doi: 10.1024/0300-9831.71.2.97.

van Kappel, A. L., J. P. Steghens, A. Zeleniuch-Jacquotte, V. Chajes, P. Toniolo, and E. Riboli. 2001. "Serum carotenoids as biomarkers of fruit and vegetable consumption in the New York Women's Health Study." *Public Health Nutr* 4 (3):829–35.

Woodall, A. A., G. Britton, and M. J. Jackson. 1997. "Carotenoids and protection of phospholipids in solution or in liposomes against oxidation by peroxyl radicals: Relationship between carotenoid structure and protective ability." *Biochim Biophys Acta* 1336 (3):575–86.

Woodall, A. A., S. W. Lee, R. J. Weesie, M. J. Jackson, and G. Britton. 1997. "Oxidation of carotenoids by free radicals: Relationship between structure and reactivity." *Biochim Biophys Acta* 1336 (1):33–42.

Woods, J. A., R. F. Bilton, and A. J. Young. 1999. "β-Carotene enhances hydrogen peroxide-induced DNA damage in human hepatocellular HepG2 Cells." *FEBS Letters* 449 (2–3):255–8. doi: 10.1016/S0014-5793(99)00450-0.

Xianquan, S., J. Shi, Y. Kakuda, and J. Yueming. 2005. "Stability of lycopene during food processing and storage." *J Med Food* 8 (4):413–22. doi: 10.1089/jmf.2005.8.413.

Yang, C. M., I. H. Lu, H. Y. Chen, and M. L. Hu. 2012. "Lycopene inhibits the proliferation of androgen-dependent human prostate tumor cells through activation of PPARgamma-LXRalpha-ABCA1 pathway." *J Nutr Biochem* 23 (1):8–17. doi: 10.1016/j.jnutbio.2010.10.006.

Yaping, Z., Q. Suping, Y. Wenli, X. Zheng, S. Hong, Y. Side, and W. Dapu. 2002. "Antioxidant activity of lycopene extracted from tomato paste towards trichloromethyl peroxyl radical CCl3O2·." *Food Chemistry* 77 (2):209–12. doi:10.1016/S0308-8146(01)00339-9.

Young, A. J., and G. M. Lowe. 2001. "Antioxidant and prooxidant properties of carotenoids." *Arch Biochem Biophys* 385 (1):20–7. doi: 10.1006/abbi.2000.2149.

Young, A. J., D. M. Phillip, and G. M. Lowe. 2004. "Carotenoid antioxidant activity." In Carotenoids in Health and Disease, edited by N.I. Krinsky, S.T. Mayne, and H. Sies, 105–126. *Marcel Dekker, Inc.*: New York.

Zaripheh, S., T. Y. Nara, M. T. Nakamura, and J. W. Erdman, Jr. 2006. "Dietary lycopene downregulates carotenoid 15,15′-monooxygenase and PPAR-gamma in selected rat tissues." *J Nutr* 136 (4):932–8.

Zhou, Y. E., M. S. Buchowski, J. Liu, D. G. Schlundt, F. A. Ukoli, W. J. Blot, and M. K. Hargreaves. 2016. "Plasma Lycopene Is Associated with Pizza and Pasta Consumption in Middle-Aged and Older African American and White Adults in the Southeastern USA in a Cross-Sectional Study." *PLoS One* 11 (9):e0161918. doi: 10.1371/journal.pone.0161918.

2 Genetic Polymorphisms Associated with Blood Concentration of Lycopene

Patrick Borel, Jean-François Landrier, and Charles Desmarchelier

CONTENTS

2.1 INTRODUCTION

Lycopene is a red pigment found mainly in tomatoes and tomato products. It belongs to the carotenoid family, which contains hundreds of molecules, and more specifically to the carotene class (i.e., non-oxygenated carotenoids). Lycopene is the carotenoid found at the highest concentration in the blood of Americans (mean 23.5–27.4 μg/dL

21

serum) (Wei, Kim, and Boudreau 2001) and in some geographical regions in Europe, for example, Ragusa and Naples in Italia (Al-Delaimy et al. 2004). The interest in this carotenoid has risen in the last decades because it has been suggested to have a protective role against the development of prostate cancer and cardiovascular diseases (Story et al. 2010; Mordente et al. 2011; Bohm 2012; Giovannucci 1999; Viuda-Martos et al. 2014; Biddle et al. 2015; Cheng et al. 2017). The mechanisms that could explain this protective role have yet to be fully elucidated but both *in vitro* and animal studies have allowed experts to suggest several hypotheses. First of all, since oxidative stress has been implicated in the etiology of these diseases, lycopene has been suggested to exert its protective effect through its antioxidant properties, which have been well characterized *in vitro* (Kelkel et al. 2011). Nevertheless, lycopene also exhibits biological activities independent of its antioxidant effects: it modulates inflammation (Gouranton et al. 2011; Marcotorchino et al. 2012; Fenni et al. 2017), reduces cholesterol absorption efficiency (Zou and Feng 2015) and there are several studies suggesting that its metabolic products also exert non-antioxidant biological effects (Mein, Lian, and Wang 2008; Gouranton et al. 2011; Aydemir et al. 2013; Gouranton et al. 2012).

Although it is likely that lycopene, together with its metabolites, has various biological effects in our body, its essentiality has not been demonstrated. Thus, it is not considered a nutrient (Hendrich et al. 1994) and there is no recommended dietary allowance for it. Nevertheless, lycopene isomers and lycopene metabolites are present in several organs, tissues, and in our blood. Blood lycopene concentration depends on several factors. The first is obviously the dietary intake of lycopene, as we are unable to synthesize it. The second depends on lycopene absorption efficiency, which is very variable because it depends on several dietary factors (Desmarchelier and Borel 2017) as well as on the ability of each individual to absorb it (Borel et al. 2015b; Bohn et al. 2017). The third is the metabolism of lycopene within the body. Indeed, it is assumed that, after absorption, lycopene is transported from the gut to the liver and then distributed to peripheral tissues. Furthermore, because it is assumed that a significant fraction of lycopene is stored in adipose tissue (Chung et al. 2009), we hypothesize that the higher the fat mass of an individual, the higher their ability to store lycopene in adipose tissue and the lower their blood concentration of lycopene. The fourth group of factors are host-related, e.g., age, gender, disease, and inflammation, which are listed in recent reviews (Desmarchelier and Borel 2017; Bohn et al. 2017). Several mechanisms involved in lycopene absorption and metabolism are governed, directly or indirectly, by proteins. It is therefore assumed that genetic variations that modulate either the activity of the proteins encoded by the genes carrying these genetic variations, or their expression, can affect lycopene absorption and/or metabolism and thus blood and tissue concentrations of lycopene (Borel 2012; Desmarchelier and Borel 2017).

In order to list the genes whose genetic variations could be suspected to affect lycopene absorption and/or lycopene metabolism, it is necessary to describe in detail all the successive metabolic steps that allow lycopene to enter the body and reach the tissues via the blood. Thus, this review starts with a detailed description of the fate of lycopene in the human body, from the food matrix in which it is ingested to extrahepatic tissues via its transit in the liver. This will allow us to identify candidate proteins, and thus candidate genes, that could carry genetic variations able to modulate blood lycopene concentration. This review then lists the genetic variations that have been associated with fasting and postprandial blood lycopene concentrations. The review finishes by listing the points to focus on in the forthcoming years to improve our knowledge on the genetic variations that modulate blood lycopene concentration.

2.2 LYCOPENE FATE IN THE GASTROINTESTINAL LUMEN DURING DIGESTION

Lycopene is insoluble in water. Thus, although it can be ingested in very different food matrices, e.g., raw tomatoes, tomato paste, or watermelon, it is assumed to transfer, at least in part, from these matrices to lipid droplets of dietary fat emulsions that are present in the gastrointestinal lumen during digestion (Borel 2003; Reboul et al. 2006; Tyssandier et al. 2003; Borel et al. 2001). This transfer, as well as the transfer of lycopene to mixed micelles, is modulated by numerous factors, e.g., food matrix, food processing, or the presence of fibres or lipids. It is beyond the scope of this review to describe the current knowledge on all these factors, but dedicated reviews can be found elsewhere (Borel 2003; Desmarchelier and Borel 2017; Bohn et al. 2015). This transfer can be facilitated by gastric and pancreatic enzymes that participate in food digestion, i.e. proteases,

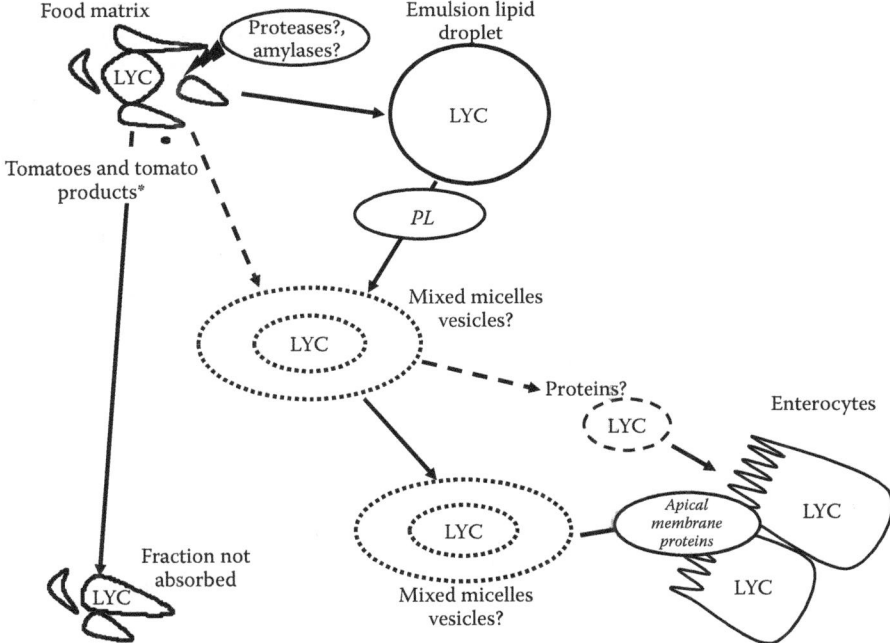

FIGURE 2.1 Proteins involved, or hypothesized to be involved, in lycopene metabolism within the lumen of the upper gastrointestinal tract. LYC: lycopene; PL: pancreatic lipase. Proteins followed by a question mark have been hypothesized to be involved because lycopene is not soluble in water and thus non-micellarized lycopene might be associated with proteins. Question marks and dotted arrows denote that this is suspected to exist but there is as yet no evidence thereof.

amylases and lipases (Figure 2.1). The genes that encode these enzymes are thus the first candidates that might affect lycopene absorption and thus blood and tissue lycopene concentrations. A fraction of lycopene is transferred from its food matrix to lipid droplets; this fraction then transfers from lipid droplets to mixed micelles in the duodenum/jejunum. Furthermore, a fraction of lycopene that remains embedded in its food matrix may also transfer to mixed micelles. Again, this transfer is assumed to be facilitated by the action of the digestive enzymes (Tyssandier, Lyan, and Borel 2001). Mixed micelles transport lycopene to the apical side of the enterocyte, where it is taken up via both passive diffusion and facilitated transport. Finally, we hypothesize that some lycopene might also associate with protein(s) or peptide(s) that also transport lycopene to the enterocyte.

2.3 APICAL UPTAKE, INTRACELLULAR METABOLISM, AND BASOLATERAL SECRETION OF LYCOPENE BY ENTEROCYTES

The mechanism by which micellarized lycopene crosses the apical membrane of the enterocyte is still not accurately known. Nevertheless, studies performed in the last decade have suggested that, contrary to what was previously assumed, lycopene uptake by the enterocyte is not only passive. This is supported by the possible saturation of absorptive mechanisms (Diwadkar-Navsariwala et al. 2003) and by the fact that several proteins, located in the apical membrane of the intestinal cell, have been involved

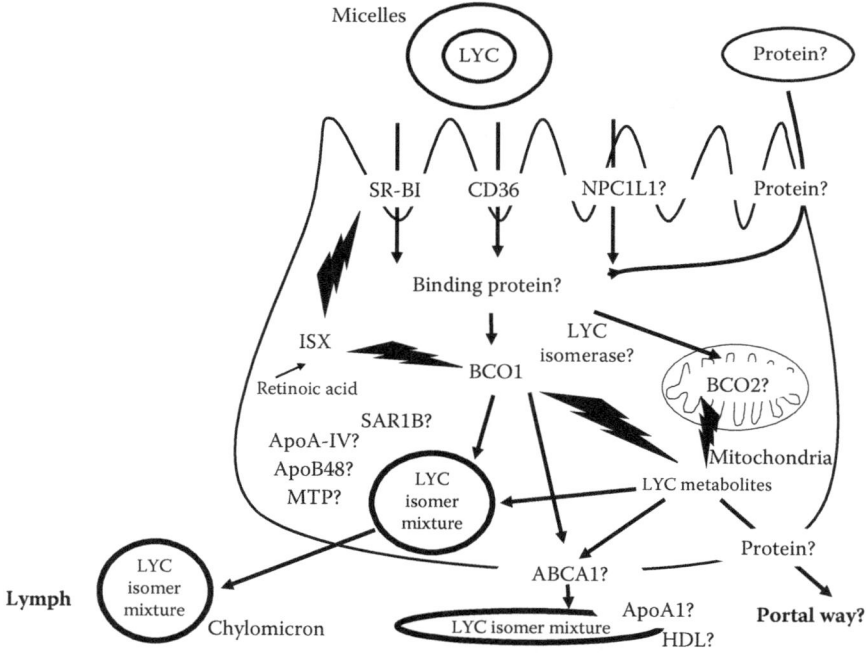

FIGURE 2.2   Proteins involved in lycopene metabolism within the enterocyte. Question marks and dotted arrows denote that this is suspected to exist but there is as yet no evidence thereof.

in lycopene uptake (Figure 2.2). These proteins are scavenger class B type I (SR-BI) (During, Dawson, and Harrison 2005; Moussa et al. 2008), which are encoded by SCARB1, CD36 molecule (CD36),—although its involvement has been demonstrated only in adipocyte (Moussa et al. 2011)—and NPC1 such as intracellular cholesterol transporter 1 (NPC1L1)—although there is conflicting data on its involvement because one study has suggested it (During, Dawson, and Harrison 2005) while another study has not (Moussa et al. 2008). It is not known whether these proteins have a direct or an indirect effect on lycopene uptake. Indeed, recent results suggest that they might indirectly modulate the apical to intracellular flux of lycopene by modulating the synthesis rate of chylomicrons (Buttet et al. 2014; Briand et al. 2016) in which lycopene is incorporated. Finally, if some lycopene is transported to the apical membrane bound to protein(s), it is not known whether it is absorbed and, if so, what is the mechanism.

After having crossed the apical membrane, lycopene must cross the polarized intestinal cell to be secreted at its basolateral side. Nothing is known about the mechanism involved but it is assumed that it compulsorily implicates proteins because lycopene is insoluble in water. Candidate proteins have been suggested (Reboul and Borel 2011). The first group are the proteins involved in the apical uptake of lycopene that display an intracellular cycle between the apical membrane and the cellular organelles. This is the case of all the above-mentioned proteins, i.e., NPC1L1, CD36, and SR-BI. Indeed, NPC1L1 has been observed in endosomes, perinuclear regions, lysosomes, and mitochondria. CD36 has been detected in the Golgi apparatus. SR-BI has been found in cytoplasmic lipid droplets and in tubulovesicular membranes. We hypothesize that lycopene binds to these proteins, or to apical membrane lipid microdomains containing these proteins, and follows their fate within the cell. The second group are intracellular enterocyte proteins that carry lipids with a broad specificity, e.g., fatty acid binding proteins (liver and intestinal FABPs, i.e., L- and I-FABP), which are both present in the intestine. The fact that a genetic association study has shown that a genetic variant in I-FABP was associated with fasting plasma lycopene concentrations (Borel et al. 2009) supports this hypothesis. The third group is composed of the enzymes responsible for carotenoid metabolism, i.e., beta-carotene oxygenase 1 (BCO1) (Lobo et al. 2012; Amengual et al. 2013) and beta-carotene oxygenase 2 (BCO2) (Hu et al. 2006). BCO1 is cytoplasmic (Lindqvist and Andersson 2002) and it is able to cleave lycopene (Dela Sena et al. 2013; Dela Sena et al. 2016) in the cytosol of mature enterocytes from the jejunum (Duszka et al. 1996). It is thus reasonable to suggest that it is also able to transport lycopene within the cell. Although BCO2 is able to cleave cis-isomers of lycopene (Hu et al. 2006), we suppose that its mitochondrial localization (Amengual et al. 2011) is not compatible with its involvement as an intracellular transporter of lycopene.

It is not known how all-trans lycopene is isomerized to cis-lycopene isomers in the enterocyte and whether there is a lycopene isomerase involved in this phenomenon, but it has been shown that there is a very efficient isomerization of lycopene in the intestinal cell (Richelle et al. 2010; Ross et al. 2011; Richelle et al. 2012). This explains why, while we eat mostly all-trans lycopene, there are mostly cis-lycopene isomers in our blood and tissues (Stahl et al. 1992).

The secretion mechanism of lycopene at the basolateral side of the intestinal cell likely depends on its cleavage in the enterocyte. Indeed, it is assumed that the parent molecules, i.e., the all-trans and cis isomers of lycopene, are incorporated in nascent chylomicrons (Borel et al. 1998), while lycopene metabolites, which are produced by BCO1 or BCO2 (Lindshield et al. 2008; Dela Sena et al. 2013; Dela Sena et al. 2016; Hu et al. 2006; Ford, Elsen, and Erdman 2013), are secreted in the portal blood. However, we suggest that some very hydrophobic lycopene metabolites, i.e., with log P > 5, might be incorporated in chylomicrons as well. The mechanisms responsible for the incorporation of lycopene in chylomicrons are barely known but it is assumed that they involve enzymes/apolipoproteins responsible for the assembly of chylomicrons, e.g., microsomal triglyceride transfer protein (MTP), apoA-IV, secretion associated Ras related GTPase 1B (SAR1B), and apoB48 (Hussain et al. 2005). It has also been suggested that the protein involved in HDL secretion, i.e., ATP-binding cassette, subfamily A, member 1 (ABCA1), might also be involved in lycopene secretion in intestinal HDL (Reboul and Borel 2011).

2.4 REGULATION OF LYCOPENE ABSORPTION AND CLEAVAGE

It is as yet unknown whether lycopene absorption and cleavage are regulated. However, the fact that SR-BI is involved in lycopene absorption (During, Dawson, and Harrison 2005; Moussa et al. 2008) and that BCO1 is involved in its cleavage (Dela Sena et al. 2013), suggests that it could be regulated by vitamin A status. Indeed, it has been demonstrated that vitamin A status regulates β-carotene absorption and cleavage efficiency via a negative feedback loop that involves an intestinal transcription factor called intestine specific homeobox (ISX) (Lobo et al. 2013; Lobo et al. 2010). ISX acts as a repressor of SCARB1 and BCO1 upon retinoic acid activation. When the intracellular concentration of retinoic acid drops, which is assumed to be the case when vitamin A intake is insufficient, ISX exerts less repressor activity towards SCARBI and BCO1 and consequently β-carotene uptake and conversion efficiency increase. This mechanism might regulate the absorption and the cleavage efficiencies of lycopene as well, although dedicated studies should be performed to verify this hypothesis.

2.5 POSTPRANDIAL BLOOD TRANSPORT OF NEWLY ABSORBED LYCOPENE FROM THE INTESTINE TO THE LIVER

As previously stated, newly absorbed lycopene is assumed to be secreted both as a mixture of all-trans and cis-isomers of the parent molecules in chylomicrons, and as lycopene metabolites in the portal blood. Concerning the second pathway, there is no information, but it can be suggested that the water-soluble metabolites of lycopene are solubilized in the portal blood and/or are incorporated in intestinal HDL and carried to the liver. Concerning the apo-B-dependent pathway, it is assumed that the all-trans and the cis isomers of lycopene are carried by chylomicrons to the liver, where they are taken up. Indeed, lycopene is apparently not significantly exchanged between lipoproteins (Romanchik, Morel, and Harrison 1995; Tyssandier et al. 2002) and thus it is assumed that it stays in the chylomicrons during their intravascular metabolism. However, these studies were performed with the all-trans isomer and it is possible that the cis-isomers are able to significantly transfer between lipoproteins or to transfer from lipoproteins to tissues during intravascular chylomicron metabolism (Zaripheh and Erdman 2005). Whatever the proportion of lycopene that is transferred to other lipoproteins or to tissues during chylomicron metabolism, it is assumed that the remaining fraction is taken up by the liver via the chylomicron-remnant receptors, i.e., the LDL-receptor (LDLR), the LDL-receptor related protein 1 (LRP1), and the heparan sulfate proteoglycans (HSPGs) (Dallinga-Thie et al. 2010).

2.6 LIVER METABOLISM AND BLOOD TRANSPORT OF LYCOPENE FROM THE LIVER TO EXTRA-HEPATIC TISSUES

Liver is the primary depot organ for lycopene (Mathews-Roth et al. 1990; Schmitz et al. 1991; Ferreira et al. 2000; Zaripheh et al. 2003; Korytko et al. 2003). Following chylomicron-remnant uptake by the liver, it is assumed that lycopene is released in hepatocytes (Figure 2.3). It is then assumed that it is transferred, by an unknown mechanism but that likely involves protein(s) (since lycopene is insoluble in the water environment of the cell), to the fat-storing cells (also called lipocytes), hepatic stellate cells, or Ito cells. Indeed, these cells, which are well known to store vitamin A (Borel and Desmarchelier 2017), are apparently also able to store lycopene (Teodoro

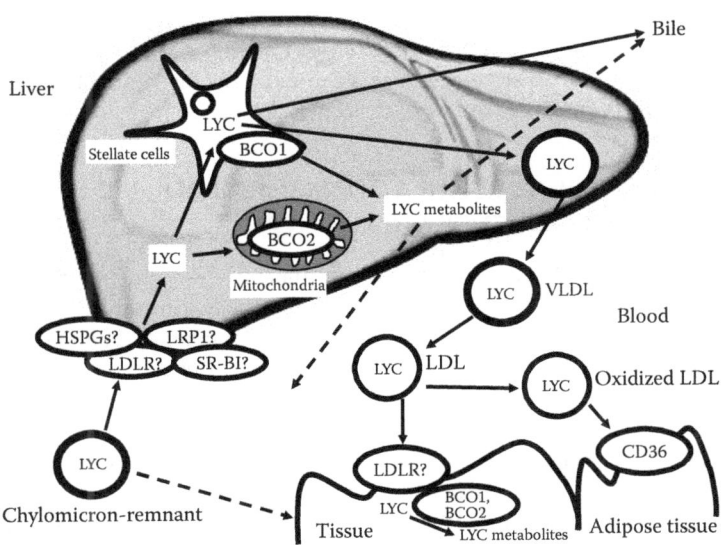

FIGURE 2.3 Proteins involved in the liver metabolism of lycopene. This figure shows current knowledge on the proteins involved in the liver metabolism of lycopene and in its distribution to peripheral tissues. Lycopene reaches the liver incorporated in chylomicron remnants following lycopene absorption. Lycopene is then mostly stored in hepatic stellate cells. The liver is assumed to be a hub of lycopene metabolism. Indeed, it is the main organ that stores lycopene and distributes it to the peripheral tissues. Question marks and dotted arrows denote that this is suspected to exist but there is as yet no evidence thereof.

et al. 2009). Again, it is not known how lycopene stored in these cells is mobilized, and which proteins are involved. Concerning its liver metabolism, it likely depends on lycopene isomerization. Indeed, the all-trans isomer is only cleaved by BCO1 (Lindshield et al. 2008; Dela Sena et al. 2013; Dela Sena et al. 2016), which is highly expressed in hepatic stellate cells (Shmarakov et al. 2010), while the 5-cis and 13-cis isomers are only cleaved by BCO2 (Hu et al. 2006; Ford, Elsen, and Erdman 2013). It is assumed that the fraction of lycopene that is not metabolized or stored in the liver is either secreted in the bile (Leo et al. 1995) or in the blood, mainly in VLDL. It is not known whether the liver also secretes lycopene metabolites in the blood. Because it is assumed that LDL, which originates from intravascular metabolism of VLDL, are responsible for the distribution of lycopene to peripheral tissues, it is hypothesized that only tissues that express the LDL-receptor can take up lycopene. However, it has been shown that CD36, which is able to bind oxidized LDL, is involved in adipose tissue uptake of lycopene (Moussa et al. 2008), and it cannot be excluded that a fraction of lycopene can reach some tissues during chylomicron metabolism (Zaripheh and Erdman 2005).

2.7   LYCOPENE METABOLISM IN EXTRA-HEPATIC TISSUES

The metabolism of lycopene in extra-hepatic tissues is barely known. It is out of the scope of this review to exhaustively list what is known in each tissue but it is necessary to talk about adipose tissue. Indeed, since it is assumed that a significant fraction of lycopene is stored in this tissue (Chung et al. 2009), we hypothesize that the ability of adipose tissue to store lycopene, or after storage to resecrete it, could have a significant impact on blood lycopene concentration.

2.8   PHYSIOLOGICAL REGULATION OF BLOOD
LYCOPENE CONCENTRATIONS

The blood concentration of lycopene can significantly vary during the period that follows a meal rich in lycopene when compared to the fasting state. For example, it increased from about 460 to about 560 nmol/l after intake, by healthy adults, of a meal that provided 19 mg of lycopene in tomato paste (Borel et al. 2016). Thus, when talking about blood lycopene concentration, it is important to specify when its concentration is measured, i.e., in the fasting period or during the postprandial period.

It is assumed that there is no direct regulation of blood lycopene concentration. In the fasting state about 75% of lycopene is transported in LDL (Romanchik, Morel, and Harrison 1995; Goulinet and Chapman 1997) but it is also found in VLDL and HDL. During the postprandial period, it is transported both in these lipoproteins and in chylomicrons (if a meal containing lycopene was ingested). Thus, we assume that blood concentration of lycopene mainly depends: 1) on the state at which the blood is collected, i.e., at fast or in the postprandial period, and when in the postprandial time; 2) on the amount of lycopene that is usually ingested by the subject, which is a key factor governing its LDL concentration; 3) on the amount of lycopene that was ingested in the meal that preceded blood sampling; 4) on the subject metabolism of lipoproteins in which lycopene is incorporated; and 5) on the ability of the subject to store/resecrete lycopene in/from their adipose tissue, which in turn depends on their body fat mass and on the metabolism of lycopene in their adipose tissue. Thus, blood lycopene concentration is likely modulated by several proteins, the main ones being those that regulate LDL (Garcia-Rios et al. 2012) and chylomicron metabolism (Dallinga-Thie et al. 2010; Perez-Martinez et al. 2010; Desmarchelier et al. 2014).

2.9   GENETIC VARIATIONS ASSOCIATED WITH BLOOD LYCOPENE CONCENTRATION

Clinical trials have systematically reported a huge interindividual variability in blood and tissue lycopene response to lycopene intake (Stahl and Sies 1992; Bowen et al. 1993; Johnson et al. 1997; O'Neill and Thurnham 1998; Diwadkar-Navsariwala et al. 2003; Wang, Edwards, and Clevidence 2013; Forman et al. 2009; Gustin et al. 2004; Borel et al. 2015b). For example, the coefficient of variation of the postprandial chylomicron lycopene response measured in 33 healthy subjects after the intake of a test meal that provided 9.7 mg lycopene in 100 g of tomato puree was 70% (Borel et al. 2015b). Genetic variations between individuals have been suggested to explain, at least in part, this phenomenon (Borel 2012; Moran, Erdman, and Clinton 2013; Desmarchelier and Borel 2017; Bohn et al. 2017). It has even been suggested that ethnically related genetic variants can influence blood carotenoid concentration (Arab et al. 2011). The use of genome-wide association studies (GWAS) to identify genetic variations associated with diseases (van der Sijde, Ng, and Fu 2014) and various phenotypes is relatively recent. These studies usually require the recruitment of thousands of subjects and are thus very expensive. This explains why there are only three GWAS dedicated to identifying genetic variations associated with blood lycopene concentration (Ferrucci et al. 2009; Zubair et al. 2015; D'Adamo et al. 2016). This also explains why there is no GWAS dedicated to identifying genetic variations associated with lycopene bioavailability. Indeed, the measurement of this phenotype requires a postprandial study, which cannot be performed easily across thousands of volunteers. In fact, there is only one candidate gene association study (CGAS) dedicated to identifying genetic variations associated with this phenotype (Borel et al. 2015b).

2.9.1   Genetic Variations Associated with Fasting Blood Lycopene Concentration

The three GWAS dedicated to identifying genetic variations associated with fasting blood lycopene concentration (Ferrucci et al. 2009; Zubair et al. 2015; D'Adamo et al. 2016) have identified single nucleotide polymorphisms (SNPs) in or near five genes (Table 2.1). In alphabetical order: BCO1 (Ferrucci et al. 2009), dehydrogenase/reductase 2 (DHRS2) (Zubair et al. 2015), SCARB1 (Zubair et al. 2015; D'Adamo et al. 2016), SET domain containing lysine methyltransferase 7 (SETD7) (D'Adamo et al. 2016) and slit guidance ligand 3 (SLIT3) (Zubair et al. 2015). It should be reminded that GWAS do not make any assumption on the genes that can affect the studied disease/phenotype and thus allow us to identify associations that were not expected to be involved in the studied disease/phenotype, in this case fasting blood lycopene concentration. This is the case for the associations found with SLIT3, STED7 and DHRS2. In fact, the associations with SLIT3 and SETD7 were not explained by the authors. SLIT3 serves as a molecular guidance cue in cellular migration, and SETD7 is one of the histone methyltransferases (HMTs) enzymes. Current knowledge on lycopene metabolism does not allow us to suggest a simple explanation of these associations. Concerning DHRS2, which encodes for an oxidoreductase, the authors only hypothesized that "it could be involved in lycopene metabolism." Thus, further studies should confirm these associations and try to explain the role of these genes on blood lycopene concentration. The associations found with SCARB1 and BCO1 are easier to explain. SCARB1 encodes for SR-BI, which is involved in lycopene absorption and tissue uptake (During, Dawson, and Harrison 2005; Moussa et al. 2008) and BCO1 encodes for beta-carotene oxygenase 1, which is able to cleave lycopene (Dela Sena et al. 2013). The GWAS have thus confirmed the key role of these proteins/ genes in lycopene metabolism.

Only two CGAS were dedicated to study associations between genetic variations in genes assumed to be involved in lycopene metabolism and fasting blood lycopene concentration (Borel et al. 2009; Borel et al. 2007) (Table 2.1). These studies focused on genes involved in lipid metabolism because lycopene is transported/stored in lipid structures in our body, e.g., lipoproteins or membranes, and it is thus assumed that its metabolism is closely related to that of lipids. These CGAS have found that SNPs in I-FABP, apoB and apoA-IV were associated with fasting blood lycopene concentration. These associations were not found in the GWAS. Thus, they likely play a minor role in this phenotype and they should be replicated in other studies. However, it should be reminded that GWAS usually lead to false negative associations, i.e., to the rejection of genetic variants that are actually associated with the studied disease/ phenotype, because of the statistical stringency used in this approach.

2.9.2 GENETIC VARIATIONS ASSOCIATED WITH BLOOD LYCOPENE RESPONSE TO DIETARY LYCOPENE

Although the identification of genetic variations associated with fasting blood lycopene concentration is of interest to better understand lycopene metabolism, the application of the results to clinical practice or dietary recommendations is otherwise not straightforward. Indeed, the highest risk factor for developing low blood lycopene concentration is usually low lycopene intake. Nevertheless, since lycopene displays a huge interindividual variability in its bioavailability, which is partly due to genetic variants, the objective to increase blood lycopene concentration should rely on identifying tailored nutritional strategies, i.e., based on the assessment of the lycopene responder phenotype of a population/individual. For example, an individual, or a group of individuals, exhibiting low blood lycopene concentrations with high capacity to absorb lycopene could be advised to increase their intake of lycopene-rich foods to increase their lycopene status. Conversely, this dietary recommendation will likely fail in subjects who are low responders to dietary lycopene and who will need supplements very rich in lycopene to improve their lycopene status.

To our knowledge, there are only two studies dedicated to identifying genetic variations associated with blood lycopene response to dietary lycopene (Table 2.1). The first
one examined the plasma lycopene response to consumption of lycopene-rich beverages
for three weeks (Wang, Edwards, and Clevidence 2013). In that study, individual respon-
siveness was associated with genetic variants in BCO1. The second study was dedicated
to identifying a combination of genetic variants associated with lycopene bioavailability,
which was evaluated by the postprandial chylomicron lycopene response to a test meal
that provided tomato puree as a source of lycopene (Borel et al. 2015b). This study showed
that lycopene bioavailability was associated with a combination of 28 SNPs in or near
16 candidate genes (Table 2.1). Seven of these genes—ABCA1, lipoprotein lipase (LPL),
insulin induced gene 2 (INSIG2), solute carrier, family 27, member 6 (SLC27A6), lipase
C, hepatic type (LIPC), CD36, and APOB—had been associated with the postprandial
chylomicron triacylglycerol response in the same group of subjects (Desmarchelier et al.
2014). This was not surprising as most newly absorbed lycopene is carried from the intestine to peripheral organs via chylomicrons, which are mainly composed of triacylglycerols. Nevertheless, four of these genes, i.e., ELVOL2, MTTP, ABCB1 and SOD2, were
specifically and significantly associated with lycopene bioavailability. Possible explana-
tions why these genes were associated with the lycopene bioavailability are discussed
in the original paper (Borel et al. 2015a). In summary, ELOVL2 is known to catalyze
the elongation of EPA (eicosapentaenoic acid) to DPA (docosapentaenoic acid) and DPA
to DHA (docosahexaenoic acid). Although lycopene is not considered to be a substrate
for this enzyme, the fact that SNPs in this gene were also associated with both lutein
(Borel et al. 2014) and β-carotene bioavailability (Borel et al. 2015a) strongly suggest that
ELOVL2 is involved, directly or indirectly, in carotenoid, and thus lycopene, metabolism. MTTP encodes for the microsomal triglyceride transfer protein, which is involved
in the packaging of triacylglycerols within the chylomicrons. Its association with lycopene bioavailability is therefore not surprising, as lycopene is incorporated into chylomicrons in the enterocyte. ABCB1 encodes for the P-glycoprotein, an ATP-dependent drug
efflux pump for xenobiotics with broad substrate specificity. Its association suggests that
this protein may participate in lycopene net absorption, possibly by effluxing a fraction
of uptaken lycopene back to the intestinal lumen. SOD2 is responsible for converting
superoxide by-products to hydrogen peroxide and diatomic oxygen. It was hypothesized
that, when this enzyme is not efficient, superoxide by-products are quenched by lycopene, leading to its degradation. Surprisingly, SNPs in genes assumed to be involved in
lycopene absorption and metabolism were not associated with lycopene bioavailability.
These SNPs were in SCARB1 and CD36, which encode for proteins involved in the
cellular uptake of lycopene (During, Dawson, and Harrison 2005; Moussa et al. 2008;
Moussa et al. 2011), and BCO1 and BCO2, which are able to cleave various isomers of
lycopene. Several hypotheses have been suggested in the original paper to explain this
lack of association (Borel et al. 2015b). It was first hypothesized that SNPs genotyped in
these genes do not result in a functionally different phenotype with regards to lycopene
bioavailability. It was also hypothesized that the association of SNPs in these genes was
weaker than that of the SNPs in the selected partial least square model. It was finally
hypothesized that some SNPs in these genes were not entered in the partial least square
regression analysis because either they were not expressed on the BeadChips, or they
were excluded from the analysis (for not following the Hardy-Weinberg equilibrium or

TABLE 2.1

Summary of SNPs Associated with Fasting Blood Lycopene Concentration or Blood Lycopene Response to Dietary Lycopene

SNP	Global MAF	Nearest Gene	Trait	Reference	Study Type
rs6564851	0.476	BCO1	FB-LYC	(Ferrucci et al. 2009)	GWAS[a]
rs74036811	0.020	DHRS2	FB-LYC	(Zubair et al. 2015)	GWAS[b]
rs7680948	0.341	SETD7	FB-LYC	(D'Adamo et al. 2016)	GWAS
rs1672879	0.269	SCARB1	FB-LYC	(Zubair et al. 2015)	GWAS
rs11057841	0.179	SCARB1	FB-LYC	(D'Adamo et al. 2016)	GWAS[a]
rs78219687	0.017	SLIT3	FB-LYC	(Zubair et al. 2015)	GWAS[b]
Apo A-IV-Ser347	0.025	APOAIV	FB-LYC	(Borel et al. 2007)	CGAS
Apo B -516	0.085	APOB	FB-LYC	(Borel et al. 2007)	CGAS
I-FABP-Thr	0.089	IFABP	FB-LYC	(Borel et al. 2009)	CGAS
rs12934922	0.357	BCO1	FB-LYC-R[c]	(Wang, Edwards, and Clevidence 2013)	CGAS
rs7501331	0.213	BCO1	FB-LYC-R[c]	(Wang, Edwards, and Clevidence 2013)	CGAS
rs1331924[e]	0.245	ABCA1	LYC-B[d]	(Borel et al. 2015b)	CGAS
rs2791952[e]	0.140	ABCA1	LYC-B	(Borel et al. 2015b)	CGAS
rs3887137[e]	0.123	ABCA1	LYC-B	(Borel et al. 2015b)	CGAS
rs4149299[e]	0.082	ABCA1	LYC-B	(Borel et al. 2015b)	CGAS
rs4149316[e]	0.122	ABCA1	LYC-B	(Borel et al. 2015b)	CGAS
rs10248420	0.347	ABCB1	LYC-B	(Borel et al. 2015b)	CGAS
rs10280101	0.145	ABCB1	LYC-B	(Borel et al. 2015b)	CGAS
rs1871744	0.499	ABCG2	LYC-B	(Borel et al. 2015b)	CGAS
rs1042031[e]	0.153	APOB	LYC-B	(Borel et al. 2015b)	CGAS
rs4112274[e]	0.224	CD36	LYC-B	(Borel et al. 2015b)	CGAS
rs3798709	0.252	ELOVL2	LYC-B	(Borel et al. 2015b)	CGAS
rs911196	0.252	ELOVL2	LYC-B	(Borel et al. 2015b)	CGAS
rs9468304	0.302	ELOVL2	LYC-B	(Borel et al. 2015b)	CGAS
rs17006621[e]	0.172	INSIG2	LYC-B	(Borel et al. 2015b)	CGAS
rs2056983	0.117	ISX	LYC-B	(Borel et al. 2015b)	CGAS
rs12914035[e]	0.096	LIPC	LYC-B	(Borel et al. 2015b)	CGAS
rs8035357[e]	0.150	LIPC	LYC-B	(Borel et al. 2015b)	CGAS
rs17482753[e]	0.088	LPL	LYC-B	(Borel et al. 2015b)	CGAS
rs7005359[e]	0.298	LPL	LYC-B	(Borel et al. 2015b)	CGAS
rs7841189[e]	0.148	LPL	LYC-B	(Borel et al. 2015b)	CGAS
rs1032355	0.251	MTTP	LYC-B	(Borel et al. 2015b)	CGAS
rs17029173	0.135	MTTP	LYC-B	(Borel et al. 2015b)	CGAS
rs745075	0.095	MTTP	LYC-B	(Borel et al. 2015b)	CGAS
rs17725246	0.254	NPC1L1	LYC-B	(Borel et al. 2015b)	CGAS
rs935933	0.139	PKD1L2	LYC-B	(Borel et al. 2015b)	CGAS
rs11197742	0.087	PNLIP	LYC-B	(Borel et al. 2015b)	CGAS

(Continued)

TABLE 2.1 (CONTINUED)
Summary of SNPs Associated with Fasting Blood Lycopene Concentration or Blood Lycopene Response to Dietary Lycopene

SNP	Global MAF	Nearest Gene	Trait	Reference	Study Type
rs10053477[e]	0.209	SLC27A6	LYC-B	(Borel et al. 2015b)	CGAS
rs9365046	0.169	SOD2	LYC-B	(Borel et al. 2015b)	CGAS

Note: The official gene symbols are those found in PubMed (available online: https://www.ncbi.nlm .nih.gov/gene/) and approved by the Hugo Gene Nomenclature Committee (available online: http:// www.genenames.org/).

Abbreviations: CGAS: candidate gene association study; GWAS: genome-wide association study; MAF: minor allele frequency (https://www.ncbi.nlm.nih.gov/projects/SNP/); FB-LYC: fasting blood lycopene concentration; FB-LYC-R: fasting blood lycopene response to dietary lycopene; LYC-B: lycopene bioavailability.

[a] Nominal association.

[b] Associations detected in African-Americans but not in Caucasian Americans of European descent.

[c] In this study, lycopene bioavailability was estimated by measuring the postprandial chylomicron lycopene response to a lycopene-rich test meal (0 to 8 h, area under the curve).

[d] Lycopene bioavailability was estimated by measuring chylomicron lycopene concentration over 8 h after consumption, by 33 healthy subjects, of a test meal that contained 100 g tomato puree. The test meal provided 9.7 mg all-trans lycopene. The coefficient of variation of the postprandial plasma chylomicron lycopene response, i.e., the area under the curve denoting 0 to 8h of lycopene concentrations, was 70%.

[e] These SNPs were associated with the variability of lycopene bioavailability, but this association was likely due to their association with chylomicron triacylglycerol response (Desmarchelier et al. 2014).

because their genetic call rate was <95%). Nevertheless, only further studies will allow us to answer whether these genes are not key genes with regard to lycopene bioavailability, at least when it is evaluated by the method used in this paper, or whether this lack of associations is due to another hypothesis.

2.10   OTHER GENETIC VARIATIONS THAT COULD BE INVOLVED IN THE BLOOD CONCENTRATION OF LYCOPENE

The low number of available studies presented in the previous paragraphs highlights the tremendous work that remains to be done to identify all the genetic variations associated with blood lycopene concentration (Table 2.1). Furthermore, there is still no study dedicated to identifying genetic variations that could modulate lycopene concentration in different tissues. It should also be remembered that, although SNPs represent ≈90% of genetic polymorphisms, other genetic variations occur in DNA, e.g., copy number variants, insertion/deletion of some base pairs, and epigenetic modifications. A genetic score that would aim to predict lycopene concentration in blood and in different tissues should therefore consider all the genetic variations that can have a significant impact on these concentrations. Finally, association studies must be performed in different populations to be sure that the associations are not specific to some ethnic groups.

In summary, there is now enough evidence to state that blood, and likely tissue, lycopene concentration, and lycopene bioavailability, are partly modulated by genetic variations in several genes. However, a lot of work remains to be done to propose combinations of genetic variations (SNPs, but also other kinds of genetic variations) that will allow us to confidently predict the concentration of lycopene in the blood or in the tissue of an individual by knowing his genotype at these variations. Yet, the potential usefulness of this area of research is exciting regarding personalized nutrition. Finally, it is important to remember that genetics only represents one of the factors that affect lycopene concentration in blood and tissues, albeit stable over the lifespan, since other factors, such as lycopene dietary intake, and factors that affect lycopene bioavailability (e.g., cooking practice), also affect this status. Thus, a prediction of lycopene concentration in blood and in various tissues should also consider these variables.

LIST OF ABBREVIATIONS

ABCA1	ATP-binding cassette, subfamily A, member 1
ABCB1	ATP-binding cassette, subfamily B (MDR/TAP), member 1
BCO1	β-carotene oxygenase 1
BCO2	β-carotene oxygenase 2
CD36	CD36 molecule
DHRS2	dehydrogenase/reductase 2
ELOVL2	ELOVL fatty acid elongase 2
HSPGs	heparan sulfate proteoglycans
INSIG2	insulin-induced gene 2
ISX	intestine specific homeobox
LDLR	LDL-receptor
L-FABP	liver-fatty acid binding protein
LIPC	lipase C, hepatic type
LPL	lipoprotein lipase
LRP1	LDL-receptor related protein 1
MTTP	gene/MTP protein (microsomal triglyceride transfer protein)
SAR1B	secretion associated Ras related GTPase 1B
SCARB1	gene/SR-BI protein (scavenger receptor, class B, member 1)
SETD7	SET domain containing lysine methyltransferase 7
SLC27A6	solute carrier, family 27, member 6
SLIT3	slit guidance ligand 3
SNPs	single nucleotide polymorphisms
SOD2	superoxide dismutase 2, mitochondrial

CONFLICT OF INTEREST

The authors declare no conflict of interest.

REFERENCES

Al-Delaimy, W. K., A. L. van Kappel, P. Ferrari, N. Slimani, J. P. Steghens, S. Bingham, I. Johansson, P. Wallstrom, K. Overvad, A. Tjonneland, T. J. Key, A. A. Welch, H. B. Bueno-de-Mesquita, P. H. Peeters, H. Boeing, J. Linseisen, F. Clavel-Chapelon, C. Guibout, C. Navarro, J. R. Quiros, D. Palli, E. Celentano, A. Trichopoulou, V. Benetou, R. Kaaks, and E. Riboli. 2004. Plasma levels of six carotenoids in nine European countries: report from the European Prospective Investigation into Cancer and Nutrition (EPIC). *Public Health Nutr.* 7 (6):713–22.

Amengual, J., G. P. Lobo, M. Golczak, H. N. Li, T. Klimova, C. L. Hoppel, A. Wyss, K. Palczewski, and J. von Lintig. 2011. A mitochondrial enzyme degrades carotenoids and protects against oxidative stress. *FASEB J.* 25 (3):948–59.

Amengual, J., M. A. Widjaja-Adhi, S. Rodriguez-Santiago, S. Hessel, M. Golczak, K. Palczewski, and J. von Lintig. 2013. Two carotenoid oxygenases contribute to mammalian provitamin A metabolism. *J. Biol. Chem.* 288 (47):34081–96.

Arab, L., M. C. Cambou, N. Craft, K. Wesseling-Perry, P. Jardack, and A. Ang. 2011. Racial differences in correlations between reported dietary intakes of carotenoids and their concentration biomarkers. *Am. J. Clin. Nutr.* 93 (5):1102–8.

Aydemir, G., Y. Kasiri, E. Birta, G. Beke, A. L. Garcia, E. M. Bartok, and R. Ruhl. 2013. Lycopene-derived bioactive retinoic acid receptors/retinoid-X receptors-activating metabolites may be relevant for lycopene's anti-cancer potential. *Mol. Nutr. Food Res.* 57 (5):739–47.

Biddle, M. J., T. A. Lennie, G. V. Bricker, R. E. Kopec, S. J. Schwartz, and D. K. Moser. 2015. Lycopene dietary intervention: A pilot study in patients with heart failure. *J. Cardiovasc. Nurs.* 30 (3):205–12.

Bohm, V. 2012. Lycopene and heart health. *Mol. Nutr. Food Res.* 56 (2):296–303.

Bohn, T., C. Desmarchelier, L. O. Dragsted, C. S. Nielsen, W. Stahl, R. Ruhl, J. Keijer, and P. Borel. 2017. Host-related factors explaining interindividual variability of carotenoid bioavailability and tissue concentrations in humans. *Mol. Nutr. Food Res.* 61 (6). doi: 10.1002/mnfr.201600685.

Bohn, T., G. J. McDougall, A. Alegria, M. Alminger, E. Arrigoni, A. M. Aura, C. Brito, A. Cilla, S. N. El, S. Karakaya, M. C. Martinez-Cuesta, and C. N. Santos. 2015. Mind the gap-deficits in our knowledge of aspects impacting the bioavailability of phytochemicals and their metabolites—A position paper focusing on carotenoids and polyphenols. *Mol. Nutr. Food Res.* 59 (7):1307–23.

Borel, P. 2003. Factors affecting intestinal absorption of highly lipophilic food microconstituents (fat-soluble vitamins, carotenoids and phytosterols). *Clin. Chem. Lab. Med.* 41 (8):979–94.

Borel, P. 2012. Genetic variations involved in interindividual variability in carotenoid status. *Mol. Nutr. Food Res.* 56:228–40.

Borel, P., and C. Desmarchelier. 2017. Genetic variations associated with vitamin A status and vitamin A bioavailability. *Nutrients* 9 (246). doi: 10.3390/nu9030246.

Borel, P., C. Desmarchelier, U. Dumont, C. Halimi, D. Lairon, D. Page, J. L. Sebedio, C. Buisson, C. Buffiere, and D. Remond. 2016. Dietary calcium impairs tomato lycopene bioavailability in healthy humans. *Br. J. Nutr.* 116 (12):2091–6.

Borel, P., C. Desmarchelier, M. Nowicki, and R. Bott. 2015a. A combination of single-nucleotide polymorphisms is associated with interindividual variability in dietary beta-carotene bioavailability in healthy men. *J. Nutr.* 145:1740–7.

Borel, P., C. Desmarchelier, M. Nowicki, and R. Bott. 2015b. Lycopene bioavailability is associated with a combination of genetic variants. *Free Radic. Biol. Med.* 83:238–44.

Borel, P., C. Desmarchelier, M. Nowicki, R. Bott, S. Morange, and N. Lesavre. 2014. Interindividual variability of lutein bioavailability in healthy men: Characterization, genetic variants involved, and relation with fasting plasma lutein concentration. *Am. J. Clin. Nutr.* 100 (1):168–75.

Borel, P., P. Grolier, N. Mekki, Y. Boirie, Y. Rochette, B. Le Roy, M. C. Alexandre-Gouabau, D. Lairon, and V. Azais-Braesco. 1998. Low and high responders to pharmacological doses of beta-carotene: Proportion in the population, mechanisms involved and consequences on beta-carotene metabolism. *J. Lipid Res.* 39 (11):2250–60.

Borel, P., M. Moussa, E. Reboul, B. Lyan, C. Defoort, S. Vincent-Baudry, M. Maillot, M. Gastaldi, M. Darmon, H. Portugal, D. Lairon, and R. Planells. 2009. Human fasting plasma concentrations of vitamin E and carotenoids, and their association with genetic variants in apo C-III, cholesteryl ester transfer protein, hepatic lipase, intestinal fatty acid binding protein and microsomal triacylglycerol transfer protein. *Br. J. Nutr.* 101:680–7.

Borel, P., M. Moussa, E. Reboul, B. Lyan, C. Defoort, S. Vincent-Baudry, M. Maillot, M. Gastaldi, M. Darmon, H. Portugal, R. Planells, and D. Lairon. 2007. Human plasma levels of vitamin E and carotenoids are associated with genetic polymorphisms in genes involved in lipid metabolism. *J. Nutr.* 137 (12):2653–9.

Borel, P., B. Pasquier, M. Armand, V. Tyssandier, P. Grolier, M. C. Alexandre-Gouabau, M. Andre, M. Senft, J. Peyrot, V. Jaussan, D. Lairon, and V. Azais-Braesco. 2001. Processing of vitamin A and E in the human gastrointestinal tract. *Am. J. Physiol. Gastrointest. Liver Physiol.* 280 (1):G95–103.

Bowen, P., V. Garg, M. Stacewiczsapuntzakis, L. Yelton, and R. S. Schreiner. 1993. Variability of serum carotenoids in reponse to controlled diets containing six servings of fruits and vegetables per day. *Ann. N. Y. Acad. Sci.* 691:241–3.

Briand, O., V. Touche, S. Colin, G. Brufau, A. Davalos, M. Schonewille, F. Bovenga, V. Carriere, J. F. de Boer, C. Dugardin, B. Riveau, V. Clavey, A. Tailleux, A. Moschetta, M. A. Lasuncion, A. K. Groen, B. Staels, and S. Lestavel. 2016. Liver X receptor regulates triglyceride absorption through intestinal down-regulation of scavenger receptor class B, type 1. *Gastroenterology* 150 (3):650–8.

Buttet, M., V. Traynard, T. T. Tran, P. Besnard, H. Poirier, and I. Niot. 2014. From fatty-acid sensing to chylomicron synthesis: role of intestinal lipid-binding proteins. *Biochimie* 96:37–47.

Cheng, H. M., G. Koutsidis, J. K. Lodge, A. Ashor, M. Siervo, and J. Lara. 2017. Tomato and lycopene supplementation and cardiovascular risk factors: A systematic review and meta-analysis. *Atherosclerosis* 257:100–8.

Chung, H. Y., A. L. Ferreira, S. Epstein, S. A. Paiva, C. Castaneda-Sceppa, and E. J. Johnson. 2009. Site-specific concentrations of carotenoids in adipose tissue: Relations with dietary and serum carotenoid concentrations in healthy adults. *Am. J. Clin. Nutr.* 90 (3):533–9.

D'Adamo, C. R., A. D'Urso, K. A. Ryan, L. M. Yerges-Armstrong, R. D. Semba, N. I. Steinle, B. D. Mitchell, A. R. Shuldiner, and P. F. McArdle. 2016. A common variant in the SETD7 gene predicts serum lycopene concentrations. *Nutrients* 8 (2):82.

Dallinga-Thie, G. M., R. Franssen, H. L. Mooij, M. E. Visser, H. C. Hassing, F. Peelman, J. J. Kastelein, M. Peterfy, and M. Nieuwdorp. 2010. The metabolism of triglyceride-rich lipoproteins revisited: New players, new insight. *Atherosclerosis* 211 (1):1–8.

dela Sena, C., S. Narayanasamy, K. M. Riedl, R. W. Curley, Jr., S. J. Schwartz, and E. H. Harrison. 2013. Substrate specificity of purified recombinant human beta-carotene 15,15′-oxygenase (BCO1). *J. Biol. Chem.* 288 (52):37094–103.

Dela Sena, C., J. Sun, S. Narayanasamy, K. M. Riedl, Y. Yuan, R. W. Curley, Jr., S. J. Schwartz, and E. H. Harrison. 2016. Substrate specificity of purified recombinant chicken beta-carotene 9′,10′-Oxygenase (BCO2). *J. Biol. Chem.* 291 (28):14609–19.

Desmarchelier, C., and P. Borel. 2017. Overview of carotenoid bioavailability determinants: From dietary factors to host genetic variations. *Trends Food. Sci. Technol.* 69:270–80.

Desmarchelier, C., J. C. Martin, R. Planells, M. Gastaldi, M. Nowicki, A. Goncalves, R. Valero, D. Lairon, and P. Borel. 2014. The postprandial chylomicron triacylglycerol response to dietary fat in healthy male adults is significantly explained by a combination of single nucleotide polymorphisms in genes involved in triacylglycerol metabolism. *J. Clin. Endocrinol. Metab.* 99 (3):E484–8.

Diwadkar-Navsariwala, V., J. A. Novotny, D. M. Gustin, J. A. Sosman, K. A. Rodvold, J. A. Crowell, M. Stacewicz-Sapuntzakis, and P. E. Bowen. 2003. A physiological pharmacokinetic model describing the disposition of lycopene in healthy men. *J. Lipid Res.* 44 (10):1927–39.

During, A., H. D. Dawson, and E. H. Harrison. 2005. Carotenoid transport is decreased and expression of the lipid transporters SR-BI, NPC1L1, and ABCA1 is downregulated in Caco-2 cells treated with ezetimibe. *J. Nutr.* 135 (10):2305–12.

Duszka, C., P. Grolier, E. M. Azim, M. C. Alexandre-Gouabau, P. Borel, and V. Azais-Braesco. 1996. Rat intestinal beta-carotene dioxygenase activity is located primarily in the cytosol of mature jejunal enterocytes. *J. Nutr.* 126 (10):2550–6.

Fenni, S., H. Hammou, J. Astier, L. Bonnet, E. Karkeni, C. Couturier, F. Tourniaire, and J. F. Landrier. 2017. Lycopene and tomato powder supplementation similarly inhibit high-fat diet induced obesity, inflammatory response and associated metabolic disorders. *Mol. Nutr. Food Res.* 61 (9). doi: 10.1002/mnfr.201601083.

Ferreira, A. L. A., K. J. Yeum, C. Liu, D. Smith, N. I. Krinsky, X. D. Wang, and R. M. Russell. 2000. Tissue distribution of lycopene in ferrets and rats after lycopene supplementation. *J. Nutr.* 130 (5):1256–60.

Ferrucci, L., J. R. Perry, A. Matteini, M. Perola, T. Tanaka, K. Silander, N. Rice, D. Melzer, A. Murray, C. Cluett, L. P. Fried, D. Albanes, A. M. Corsi, A. Cherubini, J. Guralnik, S. Bandinelli, A. Singleton, J. Virtamo, J. Walston, R. D. Semba, and T. M. Frayling. 2009. Common variation in the beta-carotene 15,15′-monooxygenase 1 gene affects Circulating levels of carotenoids: A genome-wide association study. *Am. J. Hum. Genet.* 84:123–33.

Ford, N. A., A. C. Elsen, and J. W. Erdman, Jr. 2013. Genetic ablation of carotene oxygenases and consumption of lycopene or tomato powder diets modulate carotenoid and lipid metabolism in mice. *Nutr. Res.* 33 (9):733–42.

Forman, M. R., C. B. Borkowf, M. M. Cantwell, S. Steck, A. Schatzkin, P. S. Albert, and E. Lanza. 2009. Components of variation in serum carotenoid concentrations: The Polyp Prevention Trial. *Eur. J. Clin. Nutr.* 63 (6):763–70.

Garcia-Rios, A., P. Perez-Martinez, J. Delgado-Lista, J. Lopez-Miranda, and F. Perez-Jimenez. 2012. Nutrigenetics of the lipoprotein metabolism. *Mol. Nutr. Food Res.* 56 (1):171–83.

Giovannucci, E. 1999. Tomatoes, tomato-based products, lycopene, and cancer: Review of the epidemiologic literature. *J. Natl. Cancer Inst.* 91 (4):317–31.

Goulinet, S., and M. J. Chapman. 1997. Plasma LDL and HDL subspecies are heterogenous in particle content of tocopherols and oxygenated and hydrocarbon carotenoids. Relevance to oxidative resistance and atherogenesis. *Arterioscler. Thromb. Vasc. Biol.* 17 (4):786–96.

Gouranton, E., G. Aydemir, E. Reynaud, J. Marcotorchino, C. Malezet, C. Caris-Veyrat, R. Blomhoff, J. F. Landrier, and R. Ruhl. 2012. Apo-10′-lycopenoic acid impacts adipose tissue biology via the retinoic acid receptors. *Biochim. Biophys. Acta* 1811 (12):1105–14.

Gouranton, E., C. Thabuis, C. Riollet, C. Malezet-Desmoulins, C. El Yazidi, M. J. Amiot, P. Borel, and J. F. Landrier. 2011. Lycopene inhibits proinflammatory cytokine and chemokine expression in adipose tissue. *J. Nutr. Biochem.* 22:642–8.

Gustin, D. M., K. A. Rodvold, J. A. Sosman, V. Diwadkar-Navsariwala, M. Stacewicz-Sapuntzakis, M. Viana, J. A. Crowell, J. Murray, P. Tiller, and P. E. Bowen. 2004. Single-dose pharmacokinetic study of lycopene delivered in a well-defined food-based lycopene delivery system (tomato paste-oil mixture) in healthy adult male subjects. *Cancer Epidemiol. Biomarkers Prev.* 13 (5):850–60.

Hendrich, S., K. W. Lee, X. Xu, H. J. Wang, and P. A. Murphy. 1994. Defining food components as new nutrients. *J. Nutr.* 124:S1789–S92.

Hu, K. Q., C. Liu, H. Ernst, N. I. Krinsky, R. M. Russell, and X. D. Wang. 2006. The biochemical characterization of ferret carotene-9',10'-monooxygenase catalyzing cleavage of carotenoids in vitro and in vivo. *J. Biol. Chem.* 281 (28):19327–38.

Hussain, M. M., S. Fatma, X. Pan, and J. Iqbal. 2005. Intestinal lipoprotein assembly. *Curr. Opin. Lipidol.* 16 (3):281–5.

Johnson, E. J., J. Qin, N. I. Krinsky, and R. M. Russell. 1997. Ingestion by men of a combined dose of beta-carotene and lycopene does not affect the absorption of beta-carotene but improves that of lycopene. *J. Nutr.* 127 (9):1833–7.

Kelkel, M., M. Schumacher, M. Dicato, and M. Diederich. 2011. Antioxidant and antiproliferative properties of lycopene. *Free Radic. Res.* 45 (8):925–40.

Korytko, P. J., K. A. Rodvold, J. A. Crowell, M. Stacewicz-Sapuntzakis, V. Diwadkar-Navsariwala, P. E. Bowen, W. Schalch, and B. S. Levine. 2003. Pharmacokinetics and tissue distribution of orally administered lycopene in male dogs. *J. Nutr.* 133 (9):2788–92.

Leo, M. A., S. Ahmed, S. I. Aleynik, J. H. Siegel, F. Kasmin, and C. S. Lieber. 1995. Carotenoids and tocopherols in various hepatobiliary conditions. *J. Hepatol.* 23 (5):550–6.

Lindqvist, A., and S. Andersson. 2002. Biochemical properties of purified recombinant human beta-carotene 15,15'-monooxygenase. *J. Biol. Chem.* 277 (26):23942–8.

Lindshield, B. L., J. L. King, A. Wyss, R. Goralczyk, C. H. Lu, N. A. Ford, and J. W. Erdman, Jr. 2008. Lycopene biodistribution is altered in 15,15'-carotenoid monooxygenase knockout mice. *J. Nutr.* 138 (12):2367–71.

Lobo, G. P., J. Amengual, D. Baus, R. A. Shivdasani, D. Taylor, and J. von Lintig. 2013. Genetics and diet regulate vitamin A production via the homeobox transcription factor ISX. *J. Biol. Chem.* 288:9017–27.

Lobo, G. P., J. Amengual, G. Palczewski, D. Babino, and J. von Lintig. 2012. Mammalian carotenoid-oxygenases: Key players for carotenoid function and homeostasis. *Biochim. Biophys. Acta* 1821 (1):78–87.

Lobo, G. P., S. Hessel, A. Eichinger, N. Noy, A. R. Moise, A. Wyss, K. Palczewski, and J. von Lintig. 2010. ISX is a retinoic acid-sensitive gatekeeper that controls intestinal beta,beta-carotene absorption and vitamin A production. *FASEB J.* 24 (6):1656–66.

Marcotorchino, J., B. Romier, E. Gouranton, C. Riollet, B. Gleize, C. Malezet-Desmoulins, and J. F. Landrier. 2012. Lycopene attenuates LPS-induced TNF-alpha secretion in macrophages and inflammatory markers in adipocytes exposed to macrophage-conditioned media. *Mol. Nutr. Food Res.* 56 (5):725–32.

Mathews-Roth, M. M., S. Welankiwar, P. K. Sehgal, N. C. Lausen, M. Russett, and N. I. Krinsky. 1990. Distribution of [14C]canthaxanthin and [14C]lycopene in rats and monkeys. *J. Nutr.* 120 (10):1205–13.

Mein, J. R., F. Lian, and X. D. Wang. 2008. Biological activity of lycopene metabolites: Implications for cancer prevention. *Nutr. Rev.* 66 (12):667–83.

Moran, N. E., J. W. Erdman, Jr., and S. K. Clinton. 2013. Complex interactions between dietary and genetic factors impact lycopene metabolism and distribution. *Arch. Biochem. Biophys.* 539 (2):171–80.

Mordente, A., B. Guantario, E. Meucci, A. Silvestrini, E. Lombardi, G. E. Martorana, B. Giardina, and V. Bohm. 2011. Lycopene and cardiovascular diseases: An update. *Curr. Med. Chem.* 18 (8):1146–63.

Moussa, M., E. Gouranton, B. Gleize, C. E. Yazidi, I. Niot, P. Besnard, P. Borel, and J. F. Landrier. 2011. CD36 is involved in lycopene and lutein uptake by adipocytes and adipose tissue cultures. *Mol. Nutr. Food Res.* 55:578–84.

Moussa, M., J. F. Landrier, E. Reboul, O. Ghiringhelli, C. Comera, X. Collet, K. Frohlich, V. Bohm, and P. Borel. 2008. Lycopene absorption in human intestinal cells and in mice involves scavenger receptor class B type I but not Niemann-Pick C1-like 1. *J. Nutr.* 138 (8):1432–6.

O'Neill, M. E., and D. I. Thurnham. 1998. Intestinal absorption of β-carotene, lycopene and lutein in men and women following a standard meal: Response curves in the triacylglycerol-rich lipoprotein fraction. *Br. J. Nutr.* 79:149–59.

Perez-Martinez, P., J. Delgado-Lista, F. Perez-Jimenez, and J. Lopez-Miranda. 2010. Update on genetics of postprandial lipemia. *Atheroscler. Suppl.* 11 (1):39–43.

Reboul, E., A. Berton, M. Moussa, C. Kreuzer, I. Crenon, and P. Borel. 2006. Pancreatic lipase and pancreatic lipase-related protein 2, but not pancreatic lipase-related protein 1, hydrolyze retinyl palmitate in physiological conditions. *Biochim. Biophys. Acta* 1761 (1):4–10.

Reboul, E., and P. Borel. 2011. Proteins involved in uptake, intracellular transport and basolateral secretion of fat-soluble vitamins and carotenoids by mammalian enterocytes. *Prog. Lipid Res.* 50:388–402.

Richelle, M., P. Lambelet, A. Rytz, I. Tavazzi, A. F. Mermoud, C. Juhel, P. Borel, and K. Bortlik. 2012. The proportion of lycopene isomers in human plasma is modulated by lycopene isomer profile in the meal but not by lycopene preparation. *Br. J. Nutr.* 107:1482–8.

Richelle, M., B. Sanchez, I. Tavazzi, P. Lambelet, K. Bortlik, and G. Williamson. 2010. Lycopene isomerisation takes place within enterocytes during absorption in human subjects. *Br. J. Nutr.* 103 (12):1800–7.

Romanchik, J. E., D. W. Morel, and E. H. Harrison. 1995. Distributions of carotenoids and alpha-tocopherol among lipoproteins do not change when human plasma is incubated in vitro. *J. Nutr.* 125:2610–7.

Ross, A. B., T. Vuong le, J. Ruckle, H. A. Synal, T. Schulze-Konig, K. Wertz, R. Rumbeli, R. G. Liberman, P. L. Skipper, S. R. Tannenbaum, A. Bourgeois, P. A. Guy, M. Enslen, I. L. Nielsen, S. Kochhar, M. Richelle, L. B. Fay, and G. Williamson. 2011. Lycopene bioavailability and metabolism in humans: An accelerator mass spectrometry study. *Am. J. Clin. Nutr.* 93 (6):1263–73.

Schmitz, H. H., C. L. Poor, R. B. Wellman, and J. W. J. Erdman. 1991. Concentrations of selected carotenoids and vitamin A in human liver, kidney and lung tissue. *J. Nutr.* 121:1613–21.

Shmarakov, I., M. K. Fleshman, D. N. D'Ambrosio, R. Piantedosi, K. M. Riedl, S. J. Schwartz, R. W. Curley, Jr., J. von Lintig, L. P. Rubin, E. H. Harrison, and W. S. Blaner. 2010. Hepatic stellate cells are an important cellular site for beta-carotene conversion to retinoid. *Arch. Biochem. Biophys.* 504 (1):3–10.

Stahl, W., W. Schwarz, A. R. Sundquist, and H. Sies. 1992. cis-trans isomers of lycopene and beta-carotene in human serum and tissues. *Arch. Biochem. Biophys.* 294 (1):173–7.

Stahl, W., and H. Sies. 1992. Uptake of lycopene and its geometrical isomers is greater from heat-processed than from unprocessed tomato juice in humans. *J. Nutr.* 121:2161–6.

Story, E. N., R. E. Kopec, S. J. Schwartz, and G. K. Harris. 2010. An update on the health effects of tomato lycopene. *Ann. Rev. Food Sci Technol.* 1:189–210.

Teodoro, A. J., D. Perrone, R. B. Martucci, and R. Borojevic. 2009. Lycopene isomerisation and storage in an in vitro model of murine hepatic stellate cells. *Eur. J. Nutr.* 48 (5):261–8.

Tyssandier, V., G. Choubert, P. Grolier, and P. Borel. 2002. Carotenoids, mostly the xanthophylls, exchange between plasma lipoproteins. *Int. J. Vitam. Nutr. Res.* 72 (5):300–8.

Tyssandier, V., B. Lyan, and P. Borel. 2001. Main factors governing the transfer of carotenoids from emulsion lipid droplets to micelles. *Biochim. Biophys. Acta* 1533 (3):285–92.

Tyssandier, V., E. Reboul, J. F. Dumas, C. Bouteloup-Demange, M. Armand, J. Marcand, M. Sallas, and P. Borel. 2003. Processing of vegetable-born carotenoids in the human stomach and duodenum. *Am. J. Physiol. Gastrointest. Liver Physiol.* 284 (6):G913–23.

van der Sijde, M. R., A. Ng, and J. Fu. 2014. Systems genetics: From GWAS to disease pathways. *Biochim. Biophys. Acta* 1842 (10):1903–9.

Viuda-Martos, M., E. Sanchez-Zapata, E. Sayas-Barbera, E. Sendra, J. A. Perez-Alvarez, and J. Fernandez-Lopez. 2014. Tomato and tomato byproducts. Human health benefits of lycopene and its application to meat products: A review. *Crit. Rev. Food Sci. Nutr.* 54 (8):1032–49.

Wang, T. T., A. J. Edwards, and B. A. Clevidence. 2013. Strong and weak plasma response to dietary carotenoids identified by cluster analysis and linked to beta-carotene 15,15′-monooxygenase 1 single nucleotide polymorphisms. *J. Nutr. Biochem.* 24:1538–46.

Wei, W., Y. Kim, and N. Boudreau. 2001. Association of smoking with serum and dietary levels of antioxidants in adults: NHANES III, 1988-1994. *Am. J. Public Health* 91 (2):258–64.

Zaripheh, S., T. W. Boileau, M. A. Lila, and J. W. Erdman, Jr. 2003. [14C]-lycopene and [14C]-labeled polar products are differentially distributed in tissues of F344 rats prefed lycopene. *J. Nutr.* 133 (12):4189–95.

Zaripheh, S., and J. W. Erdman, Jr. 2005. The biodistribution of a single oral dose of [14C]-lycopene in rats prefed either a control or lycopene-enriched diet. *J. Nutr.* 135 (9):2212–8.

Zou, J., and D. Feng. 2015. Lycopene reduces cholesterol absorption through the downregulation of Niemann-Pick C1-like 1 in Caco-2 cells. *Mol. Nutr. Food Res.* 59 (11):2225–30.

Zubair, N., C. Kooperberg, J. Liu, C. Di, U. Peters, and M. L. Neuhouser. 2015. Genetic variation predicts serum lycopene concentrations in a multiethnic population of postmenopausal women. *J. Nutr.* 145 (2):187–92.

3 A Possible Relationship between Tomatoes, Lycopene, and Level of High-Density Lipoprotein-Cholesterol

Koichi Aizawa, Takuro Inoue,
and Hiroyuki Suganuma

CONTENTS

3.1 INTRODUCTION

3.1.1 TOMATOES, LYCOPENE, AND CARDIOVASCULAR DISEASES

Cardiovascular diseases (CVDs) are a major cause of death in many developed countries. Hence, CVDs are a considerable problem for the health of the general population. Various studies have shown that consumption of tomatoes or lycopene may be associated with a reduction in CVD risk. Lycopene was reported to be an effective antioxidant in plasma, low-density lipoprotein (LDL), and human lymphoid cells *in vitro* [1]. Therefore, it is thought that antioxidant mechanisms may have important roles in protection against CVD. In addition, lycopene is thought to act as a hypocholesterolemic factor. *In vitro* studies by Fuhrman et al. [2] showed that lycopene can inhibit 3-hydroxy-3-methylglutaryl-CoA reductase (HMGCR) activity or increase the cellular activity of the low-density lipoprotein (LDL) receptor, which are key factors in cholesterol metabolism. Agreement with these *in vitro* studies has been observed in animal studies [3,4]. Some epidemiologic studies have also suggested that increased lycopene levels in plasma or the frequency of tomato consumption may be associated with a reduction in CVD risk. These hypotheses have been supported by some clinical trials [5–8]. However, there are likely to be multiple mechanisms through which tomatoes or lycopene can protect against CVD. More studies are needed to answer these questions.

3.1.2 IMPORTANCE OF HIGH-DENSITY LIPOPROTEIN-CHOLESTEROL (HDL-C) IN CARDIOVASCULAR EVENTS

HDL-C is commonly known as "good cholesterol" because plasma levels of HDL-C have an inverse relationship with CVD risk. It has been presumed that plasma levels of HDL-C reflect the availability of functional HDL particles with atheroprotective actions, particularly the stimulation of reverse cholesterol transport from peripheral cells to the liver [9,10]. Several epidemiologic studies have suggested an inverse correlation between plasma levels of HDL-C and CVD risk [9,10]. Barter et al. [11] reported, upon *post hoc* analyses of a study assessing the predictive value of HDL-C levels in 9770 patients, that the risk of major cardiovascular events differs significantly ($P = 0.04$) across HDL-C quintiles (Figure 3.1). Therefore, a low level of HDL-C (<40 mg/dL in men and <50 mg/dL in women) has been used as a characteristic of the metabolic syndrome by the National Cholesterol Education Program Adult Treatment Panel III [12]. It has been reported, on the basis of relatively small clinical trials or *in vitro* studies, that HDL-C may protect the endothelium, inhibit oxidation of LDL, and exert anti-inflammatory and antithrombotic effects [9]. More recently, it is thought to be more important to appreciate the functionality of HDL itself [10].

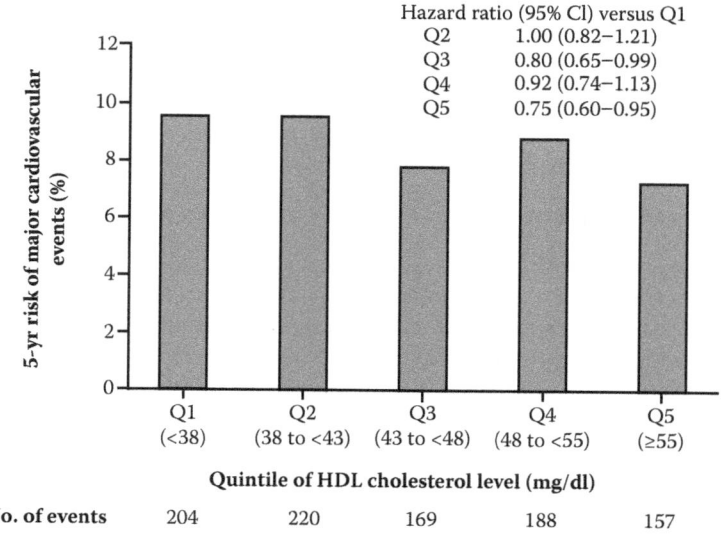

FIGURE 3.1 Multivariate analyses of the relationship between HDL-C levels at three months and the risk of major cardiovascular events [11].

3.2 FACTORS THAT MAY INFLUENCE BLOOD LEVELS OF HDL-C

A low level of HDL-C is a strong independent predictor of coronary heart disease. Low levels of HDL-C are thought to be associated with insulin resistance, increased levels of triglyceride (TG), overweight, obesity, and type-2 diabetes mellitus. Other causes are physical inactivity, cigarette smoking, very high intake of carbohydrates, and certain drugs [12]. The estimated metabolic pathway of HDL [13] is shown in Figure 3.2. HDL particles comprise lipids and proteins (including apolipoprotein A-I) and are affected by the activity of the enzyme lecithin-cholesterol acyltransferase (LCAT). Then, cholesteryl ester transfer protein (CETP) trades cholesterol ester carried on HDL for TG on very-low-density lipoprotein and LDL. These are thought to be the main factors affecting blood levels of HDL-C.

McEneny et al. [14] examined the effect of lycopene to HDL-associated inflammation in moderately overweight, middle-aged individuals. Interestingly, they found lycopene levels in serum as well as levels of HDL_2 and HDL_3 to be increased upon lycopene intervention. Furthermore, lycopene supplementation reduced CETP activity in serum, but increased LCAT activity. In conclusion, increased intake of lycopene leads to changes to levels of HDL_2 and HDL_3, which they suggested enhanced the anti-atherogenic properties of HDL_2 and HDL_3.

In addition, blood levels of HDL-C are known to be affected by low-density lipoprotein-cholesterol (LDL-C) levels. Friedewald et al. [15] presented a method for estimation of the cholesterol content of the LDL fraction. In their report, LDL-C levels were calculated according to "Friedewald's formula":

$$LDL\text{-}C = \text{total cholesterol} - HDL\text{-}C - TG/5$$

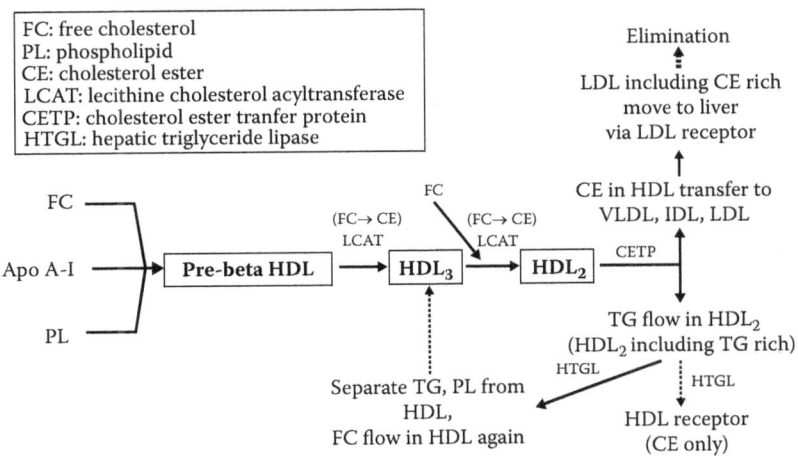

FIGURE 3.2 Estimated pathway of HDL metabolism (schematic) [13].

Friedewald's formula suggests that blood levels of HDL-C would be increased by decreased levels of LDL-C in blood. Indeed, some intervention studies in animals and humans have revealed a decrease in LDL-C levels and increase in HDL-C levels upon consumption of tomatoes or lycopene [5–8].

3.3 ANIMAL STUDIES

The anti-atherogenic effect of lycopene has been demonstrated in animals eating a high-fat diet. In such models, ingestion of tomatoes or lycopene decreased blood levels of LDL-C significantly in tandem with an increase in HDL-C levels. For example, in rabbits fed a high-fat diet, lycopene supplementation for 12 weeks decreased LDL-C levels and increased HDL-C levels in a dose-dependent manner [3]. The authors suggested that lycopene may reduce cholesterol biosynthesis through inhibition of expression of HMGCR and acyl-CoA:cholesterol O-acyltransferase. These actions would result in lower hepatic levels of cholesterol by decreased absorption of cholesterol, thereby contributing to a simultaneous increase in fecal excretion of cholesterol in lycopene-fed rabbits [3]. Similar results upon lycopene ingestion have been observed in other rabbit-based experiments [16] and in rats fed a high-fat diet [17,18].

However, a report evaluating the effect of tomatoes and/or lycopene on lipid metabolism in a normal (healthy) condition without a high-fat diet is lacking. Therefore, our research team investigated the effect of tomato consumption on the biochemical parameters and gene expression in mice fed a normal diet [19]. Mice were given normal chow and water as a control and diluted tomato juice (TJ) for six weeks. Mice had *ad libitum* access to the normal diet, water, and TJ. Then, biochemical parameters in plasma and expression of hepatic genes were analyzed. TJ ingestion increased plasma levels of HDL-C significantly (Figure 3.3e). In addition, plasma levels of glucose in the TJ group were significantly lower compared with those in the control group, but there were no notable differences in other biochemical parameters between the control group and TJ group (Figure 3.3). Microarray analyses using hepatic genes revealed down-regulation of expression of sterol regulatory element-binding proteins and up-regulation of expression of peroxisome proliferation-activated receptors. We suggested that TJ ingestion would decrease the biosynthesis of fatty acids and stimulate specific steps in the oxidation of fatty acids (data not shown). These results suggested that tomato consumption can help to maintain healthy conditions and/or reduce CVD risk.

Recently, Periago et al. [4] reported the effect of TJ consumption on the lipid profile in the plasma of rats fed a normal diet. HDL-C levels increased significantly after TJ consumption, but significant differences were not observed in levels of total cholesterol, LDL-C, very-low-density lipoprotein-cholesterol, or TG. Also, lycopene had an inhibitory effect on HMGCR. Taken together, it seems likely that intake of tomatoes or lycopene modifies lipid metabolism appropriately in a normal condition without a high-fat diet. However, the underlying mechanisms that increase HDL-C levels are mostly unknown. Hence, further studies to understand the roles of lycopene in lipid metabolism are needed.

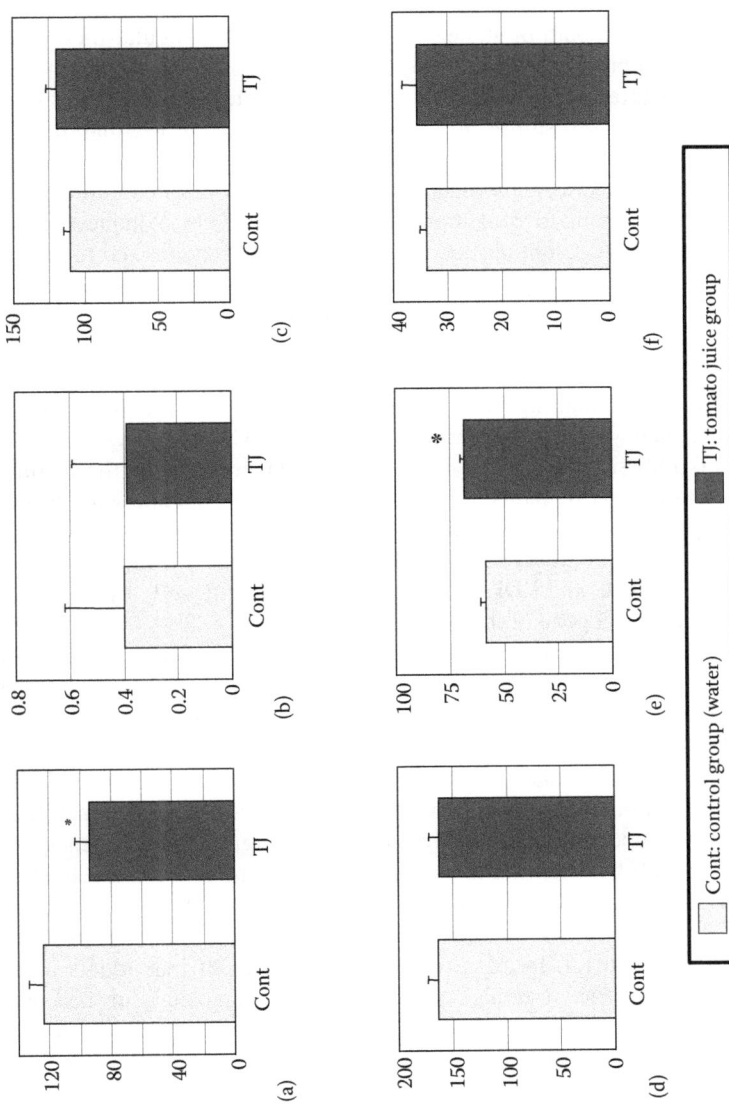

FIGURE 3.3 Changes in plasma biochemical parameters upon ingestion of water (control) or diluted tomato juice (TJ) for six weeks [19]. (a) Plasma glucose level (mg/dl); (b) plasma insulin level (ng/ml); (c) plasma total-cholesterol level (mg/dl); (d) plasma triglyceride level (mg/dl); (e) plasma high density lipoprotein (HDL) cholesterol level (mg/dl); and (f) plasma adiponectin level (μg/ml). Values are the mean ± SEM, n = 6. *Denotes differences from the control group at $P < 0.05$ according to Dunnett's multiple-comparison test.

3.4 HUMAN-INTERVENTION STUDIES

A few preliminary human-intervention studies have suggested that supplementation with tomatoes or lycopene can increase blood levels of HDL-C without alteration of LDL-C levels (Table 3.1, numbers 1–3). For example, Blum et al. [20] investigated the effects of a tomato-rich diet on the lipid profile after daily consumption of tomatoes (300 g) for one month in 98 healthy volunteers. In the regular-diet group, no changes in the lipid profile (levels of TG, total cholesterol, LDC, HDL-C) were observed before and after a one-month follow-up. In the tomato-rich-diet group, HDL-C levels were increased significantly (46.1 ± 10.6 to 53.4 ± 13.3 mg/dL, P = 0.03), but there were no significant differences in the levels of other lipids. Cuevas-Ramos et al. [21] conducted a randomized, single-blinded, controlled clinical trial to evaluate the effect of tomato consumption on HDL-C levels. Volunteers (N = 50) with low levels of HDL-C but normal levels of TG were randomized to receive 300 g of cucumber (control group) or uncooked tomatoes each day for four weeks. A significant increase in HDL-C levels was observed in the tomato group (36.5 ± 7.5 to 41.6 ± 6.9 mg/dL, P < 0.0001), but notable changes were not observed in the control group (36.8 ± 7.2 to 35.8 ± 7.3 mg/dL, P = 0.08). Levels of TG, LDL-C, and total cholesterol did not change significantly. Eva Madrid et al. [22] reported that TJ consumption for seven days increased plasma levels of HDL-C (59.13 ± 4.2 to 64.76 ± 4.6 mg/dL, P = 0.002) significantly in 17 healthy volunteers. One of the features of these studies was that participants did not suffer from hyperlipemia with normal or low levels of lipids in blood. Some reports have suggested that HDL-C levels are increased concurrently with a decrease in LDL-C levels (Table 3.1, numbers 5 and 6), or that levels of HDL-C and LDL-C are not changed significantly by consumption of tomatoes and/or lycopene (Table 3.1, numbers 7–10).

Recently, some systematic reviews and meta-analyses were carried out to evaluate evidence from intervention trials on the effect of consuming tomato products/lycopene on markers of cardiovascular function. However, a significant increasing effect on HDL-C levels by tomatoes or lycopene was not shown by one meta-analysis [23]. Interestingly, it was reported that tomato supplementation increased HDL-C levels significantly (P = 0.01) in studies from non-Western countries (Mexico, Israel, Iran, India, Korea, Greece) upon subgroup analyses, though a significant difference was not observed (P = 0.08) between tomato supplementation and HDL-C levels in all countries in that meta-analysis [24]. This observation suggests the potential effect of eating habits (including in which form tomatoes are eaten or the daily intake of tomatoes) in certain countries. In addition, the health status of individuals in the various reports used in these systematic reviews was not completely unified (e.g., heart failure, overweight, obesity, hypertension, CVD, post-menopause, metabolic syndrome, type-2 diabetes mellitus). Therefore, other types of subgroup analyses may provide some interesting new insights.

As described above, McEneny et al. [14] reported that a tomato-based diet containing lycopene and lycopene supplementation could reduce CETP activity and increase LCAT activity in serum in a human-intervention study. Their report suggested that one of the active constituents of the alternation in HDL-C levels could be lycopene. Shen et al. [25] reported that consumption of fresh tomatoes or TJ for

TABLE 3.1

Characteristics of Studies Focusing on the Relationship Between Supplementation of Tomatoes and/or Lycopene and Blood Levels of HDL-C and LDL-C

No.	Health Status	Subjects Number	Intervention	Lycopene (mg/d)	Duration	Control	Outcome	Ref.
1	Healthy	98	Tomato-rich diet	n.d.	1 mo.	Without tomato diet	HDL ↑, LDL →	[20]
2	Low HDL with normal TG	50	Uncooked tomato (300g)	n.d.	4 wk.	Cucumber (300g)	HDL ↑, LDL →	[21]
3	Healthy	17	Tomato juice	n.d.	7 d.	Non-drink	HDL ↑	[22]
4	Healthy	24	Fresh tomato Tomato juice Lycopene drink	40 40 40	6 wk.	No (baseline)	HDL ↑, LDL → HDL ↑, LDL↓ HDL →, LDL↑	[25]
5	Metabolic syndrome	27	Tomato juice	2.5 (mg/100g)	2 mo.	Non-drink	HDL ↑, LDL↓	[26]
6	Postmenopausal women	41	Tomato extract	4	6 mo.	Hormone replacement therapy (positive cont.)	HDL ↑, LDL↓	[27]
7	Grade-I hypertension	31	Tomato extract	15	8 wk.	No (baseline)	HDL →, LDL↓	[28]
8	Healthy	36	Lycopene	7	2 mo.	Placebo	HDL →, LDL↓	[29]
9	Menopausal symptoms	93	Tomato juice	44	8 wk.	No (baseline)	HDL →, LDL↓	[30]
10	Overweight	225	High tomato diet Lycopene	32–50 10	12 wk.	Low tomato diet	HDL →, LDL → HDL →, LDL →	[31]
11	Ultra-marathon runner	43	Tomato juice	n.d.	2 mo.	Other beverage	HDL →, LDL↓	[32]

Note: n.d.: not described, HDL: high-density lipoprotein cholesterol level, LDL: low-density lipoprotein cholesterol level, ↑: significantly increased (compared to control group or baseline), →: not significantly changed (compared to control group or baseline), ↓: significantly decreased (compared to control group or baseline), TG: triglyceride level.

six weeks increased blood levels of HDL-C significantly, but that consumption of a lycopene-based drink did not. This inconsistency could be due to differences in lycopene absorption. Other active constituents in tomatoes, such as phenolic compounds, may affect lipid metabolism.

The lack of robust evidence noted above hampers understanding of whether consumption of tomatoes and/or lycopene can increase blood levels of HDL-C. Further studies are required to demonstrate the role of tomatoes and/or lycopene in lipid metabolism.

3.5 CONCLUSIONS

Reports have strongly suggested that tomatoes and/or lycopene possess direct hypocholesterolemic properties. In particular, in advanced arteriosclerosis or hyperlipemia (e.g., when animals are fed a high-fat diet), lycopene consumption seems to reduce blood levels of LDL-C remarkably, and is accompanied by increases in blood levels of HDL-C. However, in the healthy/normal condition (e.g., when animals are fed a normal diet), few reports have focused on the relationship between consumption of lycopene and/or tomatoes and blood levels of lipids. Some reports have suggested that consumption of lycopene or tomatoes may increase blood levels of HDL-C alone without decreasing levels of LDL-C. Hence, lycopene or tomatoes could prevent the development of cardiovascular events by increasing blood levels of HDL-C in the normal condition. Raw tomatoes or tomato products are usually consumed by healthy people predominantly, not only individuals suffering from a cardiovascular disorder. However, very little is known about the biologic mechanisms by which lycopene works in healthy individuals. We believe that further research should focus on the relationship between lycopene intake and lipid metabolism in healthy individuals, and how lycopene can maintain this healthy condition.

REFERENCES

1. Stahl, W. and Sies, H. 1996. "Lycopene: A biologically important carotenoid for humans?" *Arch Biochem Biophys* 1;336(1):1–9.
2. Fuhrman, B., Elis, A., and Aviram, M. 1997. "Hypocholesterolemic effect of lycopene and beta-carotene is related to suppression of cholesterol synthesis and augmentation of LDL receptor activity in macrophages." *Biochem Biophys Res Commun* 233(3):658–62.
3. Verghese, M., Richardson, J. E., Boateng, J., Shackelford, L. A., Howard, C., Walker, L. T., and Chawan, C. B. 2008. "Dietary lycopene has a protective effect on cardiovascular disease in New Zealand male rabbits." *J Biol Sci* 8:268–77.
4. Periago, M. J., Martín-Pozuelo, G., González-Barrio, R., Santaella, M., Gómez, V., Vázquez, N., Navarro-González, I., and García-Alonso, J. 2016. "Effect of tomato juice consumption on the plasmatic lipid profile, hepatic HMGCR activity, and fecal short chain fatty acid content of rats." *Food Funct* 7(10):4460–7.
5. Willcox, J. K., Catignani, G. L., Lazarus, S. 2003. "Tomatoes and cardiovascular health." *Crit Rev Food Sci Nutr* 43(1):1–18.
6. Story, E. N., Kopec, R. E., Schwartz, S. J., and Harris, G. K. 2010. "An update on the health effects of tomato lycopene." *Annu Rev Food Sci Technol* 1:189–210.
7. Palozza, P., Catalano, A., Simone, R. E., Mele, M. C., and Cittadini, A. 2012. "Effect of lycopene and tomato products on cholesterol metabolism." *Ann Nutr Metab* 61(2):126–34.

8. Risseanen, T. 2006. "Lycopene and Cardiovascular Disease" in *Tomatoes, Lycopene & Human Health-Preventing Chronic Diseases*, ed. A. V. Rao, Caledonian Science Press. ISBN:0-9553565-0-4.
9. Fisher, E. A., Feig, J. E., Hewing, B., Hazen, S. L., and Smith, J. D. 2014. "High-density lipoprotein function, dysfunction, and reverse cholesterol transport." *Arterioscler Thromb Vasc Biol* 32(12):2813–20.
10. Feig, J. E., Hewing, B., Smith, J. D., Hazen, S. L., and Fisher, E. A. 2014. "High-density lipoprotein and atherosclerosis regression: Evidence from preclinical and clinical studies." *Circ Res* 114(1):205–13.
11. Barter, P., Gotto, A. M., LaRosa, J. C., Maroni, J., Szarek, M., Grundy, S. M., Kastelein, J. J., Bittner, V., and Fruchart, J. C. 2007. "Treating to New Targets Investigators. HDL cholesterol, very low levels of LDL cholesterol, and cardiovascular events." *N Engl J Med* 357(13):1301–10.
12. Executive summary of the Third Report of the National Cholesterol Education Program (NCEP) expert panel on detection, evaluation, and treatment of high blood cholesterol in adults (Adult Treatment Panel III). 2001, National institute of health, NIH Publication No. 01-3670 (May 2001).
13. Nozue, T., Inazu, A., and Mabuchi, H. 2001. "High Density Lipoprotein," in *Hyperlipidemia I, Nippon Rinsho vol. 59, suppl. 2*, ed. Suwa Y., Nippppon rinsho sya. ISBN:0047-1852.
14. McEneny, J., Wade, L., Young, I. S., Masson, L., Duthie, G., McGinty, A., McMaster, C., and Thies, F. "Lycopene intervention reduces inflammation and improves HDL functionality in moderately overweight middle-aged individuals." *J Nutr Biochem* 24(1):163–8.
15. Friedewald, W. T., Levy, R. I., and Fredrickson, D. S. 1972. "Estimation of the concentration of low-density lipoprotein cholesterol in plasma, without use of the preparative ultracentrifuge." *Clin Chem* 18(6):499–502.
16. Mulkalwar, S. A., Munjal, N. S., More, U. K., More, B., Chaudhari, A. B., and Dewda, P. R. 2012. "Effect of purified lycopene on lipid profile, antioxidant enzyme and blood glucose in hyperlipidemic rabbits." *Am J Pharm Tech Res* 2:461–70.
17. Basuny, A. M., Gaafar, A. M., and Arafat, S. M. 2009. "Tomato lycopene is a natural antioxidant and can alleviate hypercholesterolemia." *Afr J Biotechnol* 8:6627–33.
18. Jiang, W., Guo, M. H., and Hai, X. 2016. "Hepatoprotective and antioxidant effects of lycopene on non-alcoholic fatty liver disease in rat." *World J Gastroenterol* 22(46):10180–8.
19. Aizawa, K., Matsumoto, T., Inakuma, T., Ishijima, T., Nakai, Y., Abe, K., and Amano, F. 2009. "Administration of tomato and paprika beverages modifies hepatic glucose and lipid metabolism in mice: A DNA microarray analysis." *J Agric Food Chem* 57(22):10964–71.
20. Blum, A., Merei, M., Karem, A., Blum, N., Ben-Arzi, S., Wirsansky, I., and Khazim, K. 2006. "Effects of tomatoes on the lipid profile." *Clin Invest Med* 29(5):298–300.
21. Cuevas-Ramos, D., Almeda-Valdés, P., Chávez-Manzanera, E., Meza-Arana, C. E., Brito-Córdova, G., Mehta, R., Pérez-Méndez, O., Gómez-Pérez, F. J. 2013. "Effect of tomato consumption on high-density lipoprotein cholesterol level: A randomized, single-blinded, controlled clinical trial." *Diabetes Metab Syndr Obes* 6:263–73.
22. Madrid, A. E., Vásquez, Z. D., Leyton, A. F., Mandiola, C., Escobar, F. J. A. 2006. "Short-term lycopersicum esclentum consumption may increase plasma high density lipoproteins and decrease oxidative stress." *Rev Med Chile* 134:855–62.
23. Ried, K. and Fakler, P. 2011. "Protective effect of lycopene on serum cholesterol and blood pressure: Meta-analyses of intervention trials." *Maturitas* 68(4):299–310.
24. Cheng, H. M., Koutsidis, G., Lodge, J. K., Ashor, A., Siervo, M., and Lara, J. "Tomato and lycopene supplementation and cardiovascular risk factors: A systematic review and meta-analysis." *Atherosclerosis* 257:100–108.

25. Shen, Y. C., Chen, S. L., and Wang, C. K. 2007. "Contribution of tomato phenolics to antioxidation and down-regulation of blood lipids." *J Agric Food Chem* 55(16):6475–81.
26. Tsitsimpikou, C., Tsarouhas, K., Kioukia-Fougia, N., Skondra, C., Fragkiadaki, P., Papalexis, P., Stamatopoulos, P., Kaplanis, I., Hayes, A. W., Tsatsakis, A., and Rentoukas, E. 2014. "Dietary supplementation with tomato-juice in patients with metabolic syndrome: A suggestion to alleviate detrimental clinical factors." *Food Chem Toxicol* 74:9–13.
27. Misra, R., Mangi, S., Joshi, S., Mittal, S., Gupta, S. K., and Pandey, R. M. 2006. "LycoRed as an alternative to hormone replacement therapy in lowering serum lipids and oxidative stress markers: A randomized controlled clinical trial." *J Obstet Gynaecol Res* 32(3):299–304.
28. Engelhard, Y. N., Gazer, B., and Paran, E. 2006. "Natural antioxidants from tomato extract reduce blood pressure in patients with grade-1 hypertension: A double-blind, placebo-controlled pilot study." *Am Heart J* 151(1):100.
29. Gajendragadkar, P. R., Hubsch, A., Mäki-Petäjä, K. M., Serg, M., Wilkinson, I. B., and Cheriyan J. 2014. "Effects of oral lycopene supplementation on vascular function in patients with cardiovascular disease and healthy volunteers: A randomised controlled trial." *PLoS One* 9(6):e99070.
30. Hirose, A., Terauchi, M., Tamura, M., Akiyoshi, M., Owa, Y., Kato, K., and Kubota, T. 2015. "Tomato juice intake increases resting energy expenditure and improves hypertriglyceridemia in middle-aged women: An open-label, single-arm study." *Nutr J* 14:34.
31. Thies, F., Masson, L. F., Rudd, A., Vaughan, N., Tsang, C., Brittenden, J., Simpson, W. G., Duthie, S., Horgan, G. W., and Duthie, G. 2012. "Effect of a tomato-rich diet on markers of cardiovascular disease risk in moderately overweight, disease-free, middle-aged adults: A randomized controlled trial." *Am J Clin Nutr* 95(5):1013–22.
32. Samaras, A., Tsarouhas, K., Paschalidis, E., Giamouzis, G., Triposkiadis, F., Tsitsimpikou, C., Becker, A. T., Goutzourelas, N., and Kouretas, D. 2014. "Effect of a special carbohydrate-protein bar and tomato juice supplementation on oxidative stress markers and vascular endothelial dynamics in ultra-marathon runners." *Food Chem Toxicol* 69:231–6.

4 Lycopene, Tomatoes, and Cardiovascular Diseases

Volker Böhm

CONTENTS

4.1 INTRODUCTION

Cardiovascular diseases (CVDs) are the leading causes of human morbidity and mortality in developed countries. CVDs accounted for 31% (17.5 million) of all deaths globally according to WHO (World Health Organization) data from 2012 [1]. The most important risk factors are unhealthy diet, physical inactivity, tobacco use, and harmful use of alcohol. Thus, cessation of tobacco use, reduction of salt in the diet, consuming fruits and vegetables, regular physical activity, and avoiding harmful use of alcohol have been shown to reduce the risk of CVDs [1]. Changes in diet and its effect on CVD incidence have long been investigated. One recommendation is to increase consumption of fruits and vegetables, which provide a good source of various antioxidants [2].

Carotenoids—the yellow, orange, and red-colored pigments of various fruits and vegetables—are one class of compounds that have long been discussed as CVD-preventive food ingredients [3]. More than 750 carotenoids have been characterized, with approximately 50 compounds being present in the human diet. Analyzing human blood plasma samples as well as tissue samples, 10–15 carotenoids have been determined, with lycopene, α-carotene, β-carotene, lutein, zeaxanthin, and β-cryptoxanthin being the main compounds (Figure 4.1) [4,5]. Lycopene is an acyclic carotenoid with 11 conjugated double bonds, being responsible, for example, for the red colour of tomatoes. Tomatoes and tomato products are the main food sources of lycopene in our diet. In addition, pink grapefruits, papayas, guavas, rosehip products, and sea buckthorn berry products contain lycopene. As oxidation of cholesterol

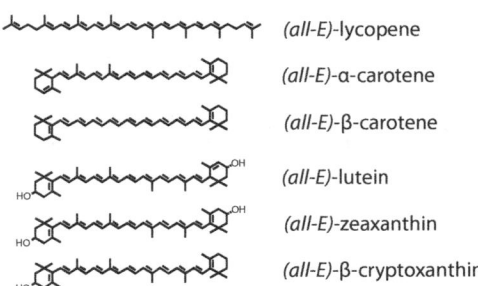

FIGURE 4.1   Main carotenoids in human plasma.

in arteries is discussed as one mechanism leading to CVDs, the antioxidant activity of carotenoids has been investigated as one preventive action of these lipophilic food ingredients [3]. In *in vitro* experiments, lycopene has been shown as a very efficient singlet oxygen quencher with approximately twice the activity of β-carotene [6]. Lycopene was also able to scavenge other reactive oxygen substances (ROS), such as superoxide radicals, peroxyl radicals, and hydroxyl radicals [6]. Our own *in vitro* investigations using different assay systems showed a high ferric-reducing activity (2.1 times higher than α-tocopherol) as well as a good peroxyl radical scavenging activity (13.3 times higher than α-tocopherol) for lycopene compared to other carotenoids [7].

4.2   LYCOPENE

Lycopene is a natural pigment, mainly present in tomato, which belongs to the carotenoid family and more precisely to the carotenes subgroup ($C_{40}H_{56}$) [4]. Like many carotenoids, lycopene is a tetraterpene. It is characterized by a symmetric and acyclic structure containing 11 linearly arranged conjugated double bonds. In the IUPAC nomenclature for carotenoids [8,9], lycopene is called ψ,ψ-carotene. In nature, lycopene is mostly present under its all-E- (or all-trans-) form (Figure 4.2), which is the most thermodynamically stable form, but some double bonds can be isomerized into Z- (or cis-) form, thus giving Z- (or cis)-lycopene isomers (Figure 4.2). Theoretically, each of the 11 carbon-carbon double bonds can isomerize, resulting in an array of mono- or poly-Z isomers of lycopene; altogether, $2^{11}=2048$ isomers are possible. But due to steric hindrance, only some ethylenic groups can be isomerized from E- to Z-form [10]. Finally, ca. 72 lycopene Z-isomers are structurally favorable and only some have been observed (Figure 4.2). One of the most common Z-lycopene isomers in nature is a *(tetra-Z)*-lycopene ((7Z,9Z,7′Z,9′Z)-lycopene; Figure 4.2) found in the tangerine tomato variety [11] and also called prolycopene.

Conversion of *all-E*-form to Z-isomers of lycopene can occur in the presence of light, exposure to heat, or by specific (bio)chemical reactions. The chemical structure of *(all-E)*-lycopene determines its physico-chemical properties. The 11 conjugated carbon-carbon double bonds form a chromophore that confers its light-absorbing properties and its bright red colour to *(all-E)*-lycopene. Its UV-visible absorption

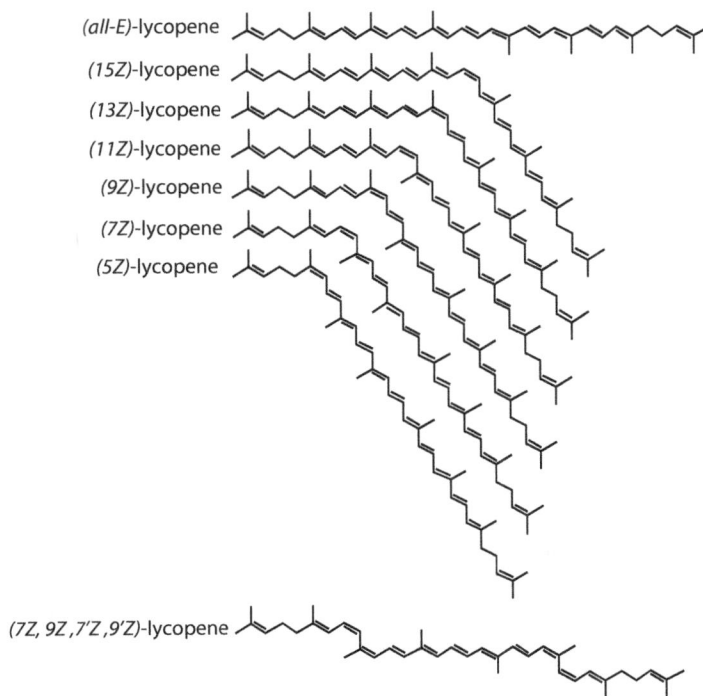

(all-E)-lycopene
(15Z)-lycopene
(13Z)-lycopene
(11Z)-lycopene
(9Z)-lycopene
(7Z)-lycopene
(5Z)-lycopene

(7Z, 9Z,7'Z,9'Z)-lycopene

FIGURE 4.2 *(all-E)*-Lycopene and some (Z)-isomers of lycopene.

spectrum is characterized by a vibrational fine structure with three maxima of absorption in the visible region (λ_{max} = 444, 470, 502 nm, % III/II = 65, ε = 184900 M^{-1}; cm^{-1} at 470 nm in light petroleum) [12]. Z-isomers of lycopene have different UV-visible characteristics. As for all carotenoids, general tendencies in comparison to *(all-E)*-lycopene are: hypsochromic shift and hypochromic effect, reduction of the vibrational fine structure, and the appearance of a new absorption band in the UV region called "cis-peak," about 142 nm below the longest-wavelength absorption maximum in the spectrum of *(all-E)*-lycopene when measured in hexane [12]. For *(poly-Z)*-lycopenes, a large hypsochromic shift can be seen (prolycopene: λ_{max} = 417, 437, 461 nm in hexane containing 2% (v/v) dichloromethane) [13].

Due to its hydrocarbon structure, *(all-E)*-lycopene is very lipophilic. It is only soluble in apolar solvents: hexane, light petroleum, benzene, chloroform, methylene chloride, and tetrahydrofurane. It is slightly soluble in acetone and ethanol and completely insoluble in water [14]. Z-isomers of lycopene are slightly more soluble in organic solvents than *(all-E)*-lycopene due to their lower tendency to form aggregates.

Because of its conjugated structure, *(all-E)*-lycopene is a fairly unstable molecule when isolated. It is highly susceptible to oxidation, and also to exposure to light, high temperatures, and extreme pH. However, in its natural matrix—i.e., tomato—lycopene was shown to be rather stable, probably protected by the matrix itself through interactions with proteins and/or membrane lipids. Degradation compounds

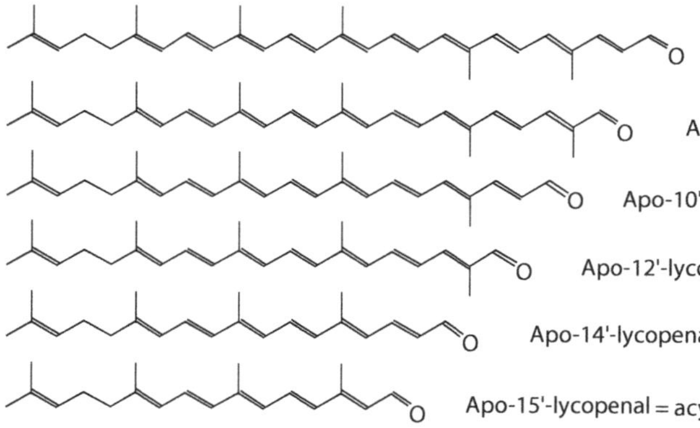

FIGURE 4.3 Chemical structures of apo-lycopenals found in human plasma by Kopec et al. [17].

of lycopene, usually found in food matrices and *in vitro* oxidation models, are issued from oxidation and/or cleavage reactions [15]. Lycopene epoxides were observed in heated tomato purée [16] and can be formed *in vitro* by chemical reaction with meta-chloro-perbenzoic acid. But lycopene epoxides are unstable molecules and usually rearrange in acidic medium to form cleavage compounds called apo-lycopenoids, with a ketone or an aldehyde ending function. Apo-lycopenoids with aldehyde ending functions, i.e., apo-lycopenals, have been detected in heated tomato products [16,17] and also in the plasma of volunteers after consumption of tomato food products [17] (Figure 4.3). Apo-lycopenoids could be natural metabolites of lycopene, but this has not yet been fully proven. Like other carotenoid oxidation products, they could have biological activities either similar and/or improved or different from those of lycopene [18–20] concerning antioxidant or cell-signalling activities [21]. For instance, in an *in vitro* model mimicking oxidative stress of dietary origin, we were able to show that long-chain apo-lycopenoids display better antioxidant properties than lycopene [22] and that the chain length and the ending function of the apo-lycopenoids can modulate their antioxidant activity and mechanisms [14]. Diapo-lycopenoids were shown to exert a higher cell-signalling effect than lycopene in the antioxidant response element (ARE) transcription system [23] and apo-10'-lycopenoic acid was proven to have an impact on adipose tissue biology via the retinoic acid receptors [24].

4.3 TOMATOES AND TOMATO PRODUCTS

Orange- and red-coloured fruits and vegetables are the main sources of dietary carotenoids. Raw tomatoes and tomato products, watermelons, guavas, grapefruits, and papayas are the richest sources of lycopene [25] (Table 4.1). Tomato products, such as juice, ketchup, paste, sauce, and soup, are especially good dietary sources of lycopene. More than 80% of the US dietary intake of lycopene is estimated to come from

TABLE 4.1

Lycopene Contents in Selected Foods

Food	Lycopene [mg/100 g]	References
Tomato, raw	0.8–3.7	[26]
Tomato juice	8–11	[27]
Tomato ketchup	15.3	[27]
Peeled tomatoes, canned	14.5	[27]
Tomato paste, concentrated	32–94	[27]
Papaya	4.3–7.5	[28]
Grapefruit	0.75	[27]
Guava	8–18	[28]
Watermelon	4.77	[27]

tomato sources, including ketchup, tomato juice, spaghetti sauce, and pizza sauce [29,30]. Moreover, dietary intake of lycopene varies greatly depending on the studied population; the average Italian consumes 14.3 mg/day of total carotenoids [31].

The bioavailable lycopene is that part of the ingested carotenoid available to the body through absorption. This bioavailability can be affected by a number of factors, including food processing and dietary composition. The amount of lycopene in fresh tomato fruits depends on different variables, such as variety, maturity, and the environmental conditions under which the fruit matured. In general, lycopene content in tomatoes is about 3 mg/100 g of wet weight [32]. Several studies about lycopene distribution in tomatoes have shown that the skin and the pericarp of tomato fruits are rich in lycopene [33,34]. This is an important aspect that must be considered during tomato processing: in fact, seeds, pericarp tissue, and skin residues are the main components of the waste during industrial treatments. So, a considerable amount of lycopene is normally discarded as tomato processing waste.

Regarding thermal stability of lycopene in tomatoes during industrial processing, a study has demonstrated that this carotenoid was stable under the conditions used. In fact, the lycopene content of tomatoes remained unchanged during the different multistep processing operations for the production of juice and paste and remained stable for up to 12 months of storage at ambient temperatures [35]. In tomato, lycopene is attached to membranes and is not released very easily. It has been shown that the process of cooking or homogenizing increases bioavailability of lycopene by its dissociation and release from the food matrix into the lipid phase of the meal [36–39]. During the cooking of tomatoes, this association of lycopene to membranes is weakened. Therefore, lycopene is better available from cooked tomatoes than from fresh tomatoes. Tomato paste and tomato puree have been shown to be more bioavailable sources of lycopene than uncooked food sources such as raw tomatoes [37,40]. In tomato purée, Sanchez-Moreno et al. [41] showed that lycopene content was not affected by pasteurization (60–85°C). Furthermore, enhanced lycopene extractability after hot air drying has been reported by Kerkhofs et al. [42].

4.4　CARDIOVASCULAR DISEASES

Cardiovascular diseases (CVDs) are the leading causes of human morbidity and mortality in developed countries. CVDs accounted for 31% (17.5 million) of all deaths globally according to WHO data from 2012 [1]. Oxidation of human LDL is a fundamental mechanism in the initiation of atherosclerosis. The presence of this modified lipoprotein in the sub-intimal space of an artery is recognized by scavenger receptors (CD36 and SRA-1). The oxidized LDLs are rapidly removed by macrophages, and if the LDL particles are rich in lipids and cholesterol, then the formation of "foam cells" is more likely to occur. Foam cells are one of the more easily recognized cells present in fatty streaks, the primary lesion identified in atherosclerosis. The altered characteristics of the macrophage means that cholesterol is retained within the arterial cell wall, and foam cells secrete growth factors that encourage the growth of smooth muscle cells in the medial layer of the artery and also secrete metalloproteinases that may destabilize the plaque, and a tissue factor that encourages platelet aggregation upon disruption of the plaque [43].

One of the initial hypotheses presented for the pathogenesis of atherosclerosis was the "response to injury hypothesis." This is considered to be a chronic inflammatory response that depends upon an initial lesion in the endothelial cell lining of the artery. Plasma proteins and LDL and modified LDL extravasate through the leaky or damaged arterial endothelial cells into the sub-intimal space. Upon transformation of the monocytes to macrophages in the sub-intimal space, the oxidized LDL particles are removed by scavenger receptors on tissue macrophages that lead to foam cell development. Damage to the endothelial cells encourages a further infiltration of monocytes and oxidized LDL particles at the sites of arterial damage. Damage to the intimal layer may also cause deposition of platelets [44], secretions from which may influence the growth of smooth muscle cells in the media layer of the artery. The "oxidative mechanism" depends upon non-modified LDL interacting with the arterial endothelial cells, and infiltrating the sub-intimal space. Oxidation can then only take place in the sub-intimal space. The oxidized species act as a chemo-attractant for monocytes by increasing the expression of monocyte chemotactic protein (MCP-1); they may also promote endothelial dysfunction acting to reduce nitric oxide production [45].

In another theory (the "inflammatory model"), the innate and adaptive arms of the immunological pathways are responsible for the development of the disease. In this model, the arterial endothelial cells express adhesion molecules for the recruitment of leukocytes and particularly monocytes, encouraging the movement of monocytes into the sub-intimal space and increasing the probability of the formation of foam cells and fatty streaks. T-cells may also enter the sub-intimal space and secrete cytokines that promote the migration and growth of smooth muscle cells initially present in the media. This can lead to plaque formation; also, the inflammatory response may be responsible for weakening of the fibrous cap that protects the plaque, which can lead to thrombotic events [46–48].

One common factor with all of the three described theories is the presence of oxidized LDL. Regarding the structure, LDL particles comprise of a single copy of apolipoprotein B100 (apoB100), phospholipids, cholesterol esters, free cholesterol, triglycerides, and lipid-soluble antioxidants such as tocopherols, β-carotene, and lycopene. The apoB100 and phospholipids are distributed towards the surface to aid stabilization of the hydrophobic core [49]. The composition of the core may vary from one particle to another and from individual to individual. This means that some individuals are present with a distribution of LDL particles that are smaller and more dense, and these particular individuals may be at greater risk of developing atherosclerosis [50,51]. There is much debate as to the nature of the oxidants that can initiate oxidative modifications to LDL particles. The modification of apoB100 is responsible for the uptake of modified LDL. Certain products from lipoperoxidation, such as nonenal, can result in structural modifications of the protein. Further possible oxidizing agents are, for example, myeloperoxidase, peroxynitrite, and metal ions [52].

4.5 HEALTH-RELATED EFFECTS

4.5.1 *IN VITRO* EXPERIMENTS

Lycopene is believed to be the most potent carotenoid antioxidant *in vitro* [6,7,53–55]. Lycopene can react with ROS in three different mechanisms: (I) by electron-transfer (ET); (II) by hydrogen atom transfer (HAT); or (III) by adduct formation [56,57]. Which type of reaction between the carotenoid and the ROS is preferred depends on the molecular structure of the carotenoid and the kind of ROS. Lycopene is able to deactivate singlet oxygen mainly by physical quenching. Only 0.05% of carotenoid activity is chemical quenching [58]. In quenching singlet oxygen ($^1O_2$) in homogeneous solution, lycopene was twice as potent compared to β-carotene and ten times more active than α-tocopherol [6,59]. Incorporated in human serum albumin as well as in dipalmitoylphosphatidylcholine membranes, lycopene and β-carotene displayed similar activity against $^1O_2$ [60,61]. Controversial results were observed in autoxidation experiments of liposomal membranes formed from cholesterol and dilinoleoylphosphatidylcholine. Similar to β-carotene, lycopene was incorporated in the inner hydrophobic part of the membranes and LDL particles due to its high lipophilicity. The X-ray experiments showed that both carotenoids disordered model membranes and stimulated lipid peroxidation of the included unsaturated fatty acids. In contrast, xanthophylls, which are located in membranes like rivets, and their polar end groups were oriented to the water phase, and especially astaxanthin decreased the levels of lipid hydroperoxides significantly [62].

In contrast to β-carotene, which adds superoxide radicals $\left(O_2^{\bullet-}\right)$, lycopene reacts with $O_2^{\bullet-}$ by electron-transfer [63]. However, the computational studies of Galano et al. showed that both carotenes should be less active in quenching $O_2^{\bullet-}$ than xanthophylls, especially carbonyl-containing ones like canthaxanthin and astaxanthin [64].

Besides singlet oxygen and superoxide, hydrogen peroxide (H_2O_2) plays an important role as a non-radical oxidant in the human vascular system. Most of the releasing H_2O_2 is reduced by erythrocyte-standing catalases. Aside, exogenous antioxidants in the blood stream can be helpful to support the protective action of these enzymes to prevent damages of blood lipids caused by H_2O_2. Lu et al. [65] showed that lycopene reacts under acidic conditions with H_2O_2, forming lycopene-1,2-epoxide and 2,6-cyclolycopene-1,5-epoxide. In human serum and milk, 2,6-cyclolycopene-1,5-epoxide is present, as was first described by [66] and could be a possible metabolite of the reaction of lycopene with H_2O_2.

Chain-breaking antioxidants are able to inhibit the propagation phase of lipid peroxidation by scavenging peroxyl radicals, amongst others. To analyze the ability of compounds to act as chain-breaking antioxidants *in vitro*, peroxyl radicals were most often generated by thermal degradation of diazo compounds such as 2,2'-azo-bis(2,4-dimethylvaleronitrile) (AMVN), 2,2'-azo-bis(isobutyronitrile) (AIBN) and 2,2'-azo-bis(2-methylamidinopropane) (AAPH). The activity of plasma antioxidants (e.g., lycopene) to deactivate peroxyl radicals depends on the location of the peroxyl radicals and the antioxidants and therefore on their polarity. In a plasma model, lycopene was the strongest carotenoid against peroxyl radicals formed by thermal degradation of the more hydrophilic azo-compound AAPH, however less antioxidant than α-tocopherol and hydrophilic antioxidants. In contrast, against an attack of peroxyl radicals, initiated by the apolar MeO-AMVN, carotenoids action was comparable to that of α-tocopherol [67]. Our own studies using AAPH in a homogeneous model system showed lycopene being approximately ten times more active than α-tocopherol but less active than the more polar xanthophylls, which possibly showed higher activity than lycopene or β-carotene due to a better availability by polar AAPH radicals [7]. The carbonyl-containing xanthophylls with their highly unsaturated polyene chain were especially effective in scavenging peroxyl radicals formed from AAPH, which supports the theoretical investigations of [68]. Lycopene was the most antioxidant carotenoid, not only in homogeneous solution, but also in inhibiting oxidation of multilamellar liposomes by AMVN radicals, measured by the formation of TBARS. It was twice as active as α-carotene and three times more active than β-carotene [69]. However, by using egg yolk phosphatidylcholine-based liposomes, lycopene was destroyed much faster than xanthophylls and α-tocopherol by using AAPH and AMVN as radical generators, but was the least effective antioxidant in inhibition of lipid peroxidation due to its orientation in the inner core of the micelles [54,70].

To assess the antioxidant activity of hydrophilic as well as lipophilic compounds, such as lycopene, various high-throughput methods were developed using different types of synthetic oxidants such as ABTS and DPPH. Using the synthetic 2,2'-azino-bis(3-ethylbenzothiazolin-6-sulfonic acid) radical cation (ABTS•+) several times, lycopene was observed to be the most antioxidant carotenoid and two to four times more active than α-tocopherol [7,53,71]. The DPPH assay is a very popular method of analyzing antioxidant activity, especially of alcohol-soluble compounds such as polyphenols, vitamin C, and extracts and oils of fruits, vegetables, and spices. In contrast, the activity of lycopene and carotenoids in general in scavenging the synthetic RNS (Reactive Nitrogen Species) 2,2-diphenyl-1-picrylhydrazyl radical (DPPH•) is negligible till not present. In our own studies, carotenes and xanthophylls did not react with DPPH. Only α-tocopherol, which was used as reference, showed activity [7]. The studies of [72] showed that lycopene did not act in a concentration-dependent relation with DPPH. With increasing concentrations of the carotenoid, the DPPH scavenging rate decreased. Similar results were obtained for β-carotene [72].

Transitions metals such as $Fe^{2+/3+}$ and $Cu^{+/2+}$ play an important role in the initiation of lipid peroxidation and the progression of lipid peroxidation-associated diseases. Iron could be released in the GIT from heme-containing proteins, especially in a diet rich in meat, and can oxidize lipids, which could be absorbed thereafter [73]. The ferric-reducing antioxidant power (FRAP) assay is a fast-acting method to analyze the activity of antioxidants to reduce ferric ions [74]. In our recent studies, we showed that lycopene was able to reduce ferric ions, which were incorporated in an organic complex, with similar activity like the xanthophylls lutein and zeaxanthin, and twice as active as the antioxidant vitamins α-tocopherol and ascorbic acid. In contrast, in several studies the ionone ring-containing carotenes (α- and β-carotene) did not show FRAP activity [7,71,75,76]. Additionally, similar relationships between lycopene, α-tocopherol, and ascorbic acid were observed in the cupric ion-reducing antioxidant capacity (CUPRAC) assay, which measures the activity of compounds to reduce copper(II), based on principles similar to the FRAP assay [71]. From these results, it can be hypothesized that lycopene is a strong transition metal-reducing agent.

Lycopene can also function as a potent scavenger of hypochlorous acid (HOCl). HOCl is generated enzymatically by myeloperoxidase (MPO), a neutrophil-derived heme peroxidase, which uses H_2O_2 to catalyze a two-electron oxidation of chloride. HOCl is implicated as a contributing factor in a number of pathological conditions, including inflammatory diseases and atherosclerosis [77,78]. In lipids, the major sites of attack by HOCl are the double bonds of unsaturated fatty acids or cholesterol, leading to peroxidation or chlorohydrin formation [79,80]. In addition, HOCl can react with other compounds to form other ROS, such as ˙OH and $O_2^{\bullet-}$ [81]. These ROS can cause further cellular damages. Recently, Pennathur et al. showed in homogeneous solution that lycopene scavenges HOCl in a dose-dependent manner [82]. A wide range of possible lycopene degradation products were evaluated using LC-MS, such as apo-lycopenals and carotendials. In a postprandial model of oxidative stress, physiological concentrations of lycopene inhibited the lipid peroxidation activity of metmyoglobin, a hemeprotein, against linoleic acid under mildly acidic conditions (pH 5.6).

Lycopene reacts also most effectively with nitrogen dioxide radical $\left(NO_2^{\bullet}\right)$ compared to other carotenoids. The order of reactivity was found to be: lycopene > β-carotene > β-cryptoxanthin > lutein/zeaxanthin > canthaxanthin = astaxanthin [83]. Consequently, the sequence of reactivity was closely related to the reaction with the synthetic RNS-dye ABTS˙+ [7,54] due to the fact that the reaction of both nitrogen-based radicals with carotenoids is based on electron-transfer.

But lycopene is not only described as a scavenger of ROS, RNS, and ROCl; lycopene is also able to deactivate sulphur-based radicals on a higher rate than diet-relevant xanthophylls. Experiments with *in vivo* relevant glutathione-thiyl (G-RS˙) and thiyl-sulfonyl (RSOO˙), which can be formed by reaction of thiyl with molecular oxygen, showed that lycopene reacts 2.5 times faster than lutein, zeaxanthin, canthaxanthin, and astaxanthin, by ET-mechanism or adduct formation [83] due to its larger system of conjugated double bonds.

Carotenoids are susceptible to isomerisation and oxidation. Heat, light, and acids can promote isomerization of *(all-E)*-carotenoids to their Z-forms, resulting in a loss of color [84,85]. Non-enzymatic oxidative degradation is the main cause of extensive loss of carotenoids. It depends on the availability of oxygen and could be stimulated by light, metal ions, and lipid hydroperoxides [86,87]. Furthermore, oxidizing enzymes can lead to carotenoid oxidation and degradation. Undesirable degradation of lycopene not only affects the sensory quality of the final products, but also the health benefits of tomato-based foods for the human body [85]. In model systems, lycopene was the carotenoid most susceptible to oxidation initiated by high temperature. The degradation of lycopene runs twice faster than that of β-carotene and lutein. The contribution of Z-isomers of lycopene in processed tomato-based products should not be underestimated. Whereas the *(all-E)*-form is the main isomer of lycopene in raw tomatoes (up to 96% of total lycopene) the proportion of Z-isomer was 4% to 27% in various tomato-based products such as spaghetti sauce [88]. Furthermore, Z-isomers of lycopene contribute to over 50% to total lycopene content in human plasma, with (5Z)-, (9Z)-, (13Z)- and (15Z)-lycopene as the predominant mono-(Z)-forms [88].

In simple antioxidant activity assays, the most important lycopene Z-isomers displayed the same oxidant reducing activities as the *(all-E)*-form. (5Z)-, (9Z)-, (13Z)-lycopene as well as the tetra-Z-isomer found in tangerine tomato varieties were similarly antioxidant in reducing ferric ions in the FRAP assay and in bleaching the ABTS$^{\cdot+}$ in the TEAC assay [89]. In a postprandial model of oxidative stress, the (5Z)-isomer of lycopene was more active than the *(all-E)*-form, and 50% more active than β-carotene and three times more active than α-tocopherol [89]. (5Z)-Lycopene is the predominant Z-isomer of lycopene in human plasma with a contribution of approximately one-third of the total lycopene content [90]. In the more complex studies of [91], the activity to inhibit oxidative stress and apoptosis of THP-1 cells induced by 7-ketocholesterol of the (5Z)-isomer of lycopene was not significantly different to the activity shown by the *(all-E)*-isomer.

4.5.2 *In vitro* Cell Studies

Recently, Tang et al. reported that lycopene in concentrations of 0.2–20 μM was able to protect ECV304 endothelial cells against oxidative attacks by H_2O_2, measured by reduced malondialdehyde (MDA) levels compared to a control group. In the lycopene group, the cell viability increased and the level of apoptotic cells decreased significantly [92]. TNF_α-induced intercellular adhesion molecule-1 (ICAM-1) expression in HUVECs was inhibited by lycopene in human umbilical endothelial cells (HUVECs), whereas cyclooxygenase-2 (COX-2) and platelet-endothelial cell adhesion molecule (PECAM-1) expression were not affected [93]. A further analysis indicated that lycopene inhibited TNF_α-induced NF-κB-DNA, but not AP1-DNA complex-formation. In contrast, lycopene did not affect TNF_α-induced p38 and extracellular matrix-regulated kineses 1 and 2 (ERK1/2) phosphorylation and interferon-γ (IFN-γ)-induced signalling—suggesting that lycopene primarily affects the TNF_α-induced NF-κB signalling pathway [93]. In a functional study, lycopene dose-dependently attenuated monocyte adhesion to endothelial monolayers but not that adhesion to the extracellular matrix. Additionally, lycopene down-regulated the expressions of p53 and caspase-3 mRNA induced by H_2O_2 on a similar level as probucol, which is known as an effective antioxidant drug against CVDs [94]. These findings suggest that lycopene may act as an antiatherogenic agent by a mechanism involving, at least in part, an antioxidant mechanism [95].

Experiments with the human macrophage cell line THP-1 showed an effective inhibition of oxidative stress and apoptosis induced by 7-ketocholesterol (7-KC), an oxysterol with pathophysiological relevance in atherogenesis [91]. Physiologically relevant doses (0.5–2 µM) of *(all-E)*-lycopene and its (5Z)-isomer significantly reduced the increase in ROS production and in 8-hydroxydeoxyguanosine (8-OHdG) formation induced by the oxysterol in a dose-dependent manner. Additionally, the carotenoid strongly prevented the increase of NOX-4, hsp70, and hsp90 expressions, as well as the phosphorylation of the redox-sensitive p38, JNK, and ERK1/2 induced by 7-KC. Lycopene inhibited 7-KC-induced apoptosis by limiting caspase-3 activation. Summarized, lycopene was more effective than β-carotene in counteracting the dangerous effects of 7-KC in human macrophages. The authors suggested that lycopene is a potential antiatherogenic agent to prevent oxidative stress and apoptosis of human macrophages [91]. Furthermore, lycopene in the above-mentioned concentrations reduced the intracellular content of total cholesterol in the THP-1 macrophages by reduction of the expression of 3-hydroxy-3-methylglutaryl-coenzym A (HMG-CoA) reductase [96]. Due to hypercholesterolemia as one risk factor for atherosclerosis, this implies a potential role of lycopene in decreasing foam cell formation and therefore in the risk reduction of CVDs.

Lo et al. examined the effect of lycopene on smooth muscle cells (SMC), whose accumulation in the sub-endothelial tissue is connected to CVDs. They found that pre-incubation of platelet-derived growth factor (PDGF)-BB with lycopene resulted in a marked inhibition of PDGF-BB-induced PDGF receptor-beta (PDGFR-β), PLCγ, and ERK1/2 phosphorylation in rat A10 SMCs and primary cultured aortic SMC. However, lycopene did not influence EGF-induced ERK1/2 phosphorylation. Further analysis indicated that lycopene could directly bind PDGF-BB and inhibit PDGF-BB-SMC interaction. Lycopene inhibited PDGF-BB-induced SMC proliferation and migration at concentrations ranging from 2–10 µM [97].

Lycopene concentration-dependently (2–12 µM) inhibited collagen, ADP, or arachidonic acid induced platelet aggregation in human platelets. The phosphorylation was markedly inhibited by lycopene in phosphorus-32-labeled platelets. These results indicated that the antiplatelet activity of lycopene may involve the inhibition of the activation of phospholipase C, and lycopene also activated the formations of cyclic GMP/nitrate in human platelets, resulting in the inhibition of platelet aggregation. The results may imply that tomato-based foods are especially beneficial in the prevention of platelet aggregation and thrombosis [98].

4.5.3 HUMAN INTERVENTION TRIALS

In contrast to some epidemiological studies, none of the systemic markers, such as inflammatory markers, markers of insulin resistance, and sensitivity, changed significantly in the plasma after a dietary intervention in the study by Thiess et al. [99]. After a four-week run-in period with a low-tomato diet, a moderately overweight group of 225 volunteers (94 men and 131 women) aged 40–65 years was randomly assigned into one of three dietary intervention groups and asked to consume a control diet low in tomato-based food, a high-tomato-based diet, or a control diet supplemented with lycopene capsules (10 mg/d) for 12 weeks. During the intervention, blood pressure, weight, and arterial stiffness were measured. Dietary intake was also determined. The study resulted in absent changes of lipid concentrations and arterial stiffness, though carotenoid profile changed and total concentrations increased [99]. These data indicate that a relatively high daily consumption of tomato-based products (equivalent to 32–50 mg lycopene/d) or lycopene supplements (10 mg/d) is ineffective at reducing conventional CVD risk markers in short-term studies.

Promising results were achieved in several recently done human intervention studies. Within a 12-week intervention study (54 moderately overweight individuals aged 40–65 years, BMI: 18.5–35) in the subgroup (n = 18) with 70 mg lycopene per week, comprised in a supplement, significantly decreased serum amyloid A levels were determined, showing reduced inflammation [100]. Uptake of an optimized soup (7.6 mg lycopene/d) for four weeks (69 subjects aged 30 ± 10 years within two groups [optimized soup vs. control soup]) resulted in significantly reduced levels of serum oxLDL (oxidized LDL cholesterol) [101]. Daily consumption of 160 g of tomato sauce high in lycopene (27.2 mg/d) for four weeks (30 healthy subjects aged 39 ± six years, BMI: 24.5 ± 3.3 kg/m^2) also induced a significant reduction in oxLDL levels in plasma [102]. Uptake of 7 mg lycopene per day (as a supplement) for two months improved endothelium-dependent vasodilatation by 53% in statin-treated patients (24 patients aged 67 ± six years, with stable cardiovascular diseases, BMI: 28.6 ± 3.3 kg/m^2) [103]. Daily consumption of 330 ml of tomato juice (37.0 mg lycopene/d) for 20 days (32 overweight or obese female subjects aged 25.2 ± 0.6 years, BMI: 29.4 ± 0.2 kg/m^2) significantly increased total antioxidant capacity in plasma as well as activities of superoxide dismutase, catalase, and glutathione peroxidase in erythrocytes; the level of malondialdehyde in serum was significantly decreased [104]. A diet high in processed tomato products (ca. 200 g/d: ca. 228 mg lycopene/d) for six weeks (30 overweight or obese subjects aged 45.0 ± 14.5 years, BMI: 29.9 ± 4.6 kg/m^2) led to a significant reduction in diastolic blood pressure (75.7 ± 1.5 to 72.5 ± 1.3 mmHg) while lipid status parameters, insulin concentration, and inflammation markers remained unchanged [105]. Daily consumption of 400 ml tomato juice (44 mg lycopene/d) for eight weeks (93 women aged 40–60 years, BMI: 22.2 ± 3.2 kg/m^2) led to a significantly improved menopausal symptom score, perhaps partly explained by the antioxidative effect of lycopene. In the subgroup of women (n = 22) with high levels of triglycerides at baseline, the levels decreased significantly after four and eight weeks of intervention [106].

4.5.4 Concluding Remarks

Lycopene was shown as the most effective antioxidant, compared to other carotenoids, in several *in vitro* experiments. This carotenoid was able to counteract reactive oxygen species as well as reactive nitrogen species. In cell studies, lycopene in concentrations of 0.2–20 μM protected endothelial cells against oxidative attacks. In experiments with human macrophages, lycopene effectively inhibited oxidative stress and apoptosis. Lycopene concentration-dependently (2–12 μM) inhibited agonist induced platelet aggregation in human platelets. All these *in vitro* results led to the hope that lycopene might be a preventive food ingredient regarding cardiovascular diseases. However, until now, human intervention trials presented controversial results. While some studies showed promising effects of lycopene and mostly better when derived from tomato products, others did not show any effect. Lycopene metabolites, formed in the human body, are perhaps the more important compounds compared to the native carotenoid. As lycopene is only one ingredient of tomatoes and tomato products, the more complex food matrix with other secondary plant products as well as with vitamins and minerals might be responsible for some effects. Nonetheless, antioxidative and anti-inflammatory activities of lycopene as a partial explanation of the preventive effects of tomato products have to be further investigated in the future.

REFERENCES

1. World Health Organization. 2016. Cardiovascular diseases, Fact sheet. www.who.int /mediacentre/factsheets/fs317/eng/ (accessed February 6, 2017).
2. Ruxton, C. H. S., Gardner, E. J., and Walker, D. 2006. "Can pure fruit and vegetable juices protect against cancer and cardiovascular disease too? A review of the evidence." *Int J Food Sci Nutr* 57:249–72.
3. Voutilainen, S., Nurmi, T., Mursu, J., and Rissanen, T. H. 2006. "Carotenoids and cardiovascular health." *Am J Clin Nutr* 83:1265–71.
4. Britton, G., Liaaen-Jensen, S., and Pfander, H. 2004. *Carotenoids – Handbook*, Birkhäuser Verlag, Basel.
5. Maiani, G., Periago Castón, M. J., Catasta, G., Toti, E., Goñi Cambrodón, I., Bysted, A., Granado-Lorencio, F., Olmedilla-Alonso, B., Knuthsen, P., Valoti, M., Böhm, V., Mayer-Miebach, E., Behsnilian, D., and Schlemmer, U. 2009. "Carotenoids: Actual knowledge on food sources, intakes, stability and bioavailability and their protective role in humans." *Mol Nutr Food Res* 53:S194–S218.
6. Di Mascio, P., Kaiser, S., and Sies, H. 1989. "Lycopene as the most efficient biological carotenoid singlet oxygen quencher." *Arch Biochem Biophys* 274:532–8.
7. Müller, L., Fröhlich, K., and Böhm, V. 2011. "Comparative antioxidant activities of carotenoids measured by ferric reducing antioxidant power (FRAP), ABTS bleaching assay (aTEAC), DPPH assay and peroxyl radical scavenging assay." *Food Chem* 129:139–48.
8. IUPAC. 1971. "IUPAC Commission on the Nomenclature of Organic Chemistry and IUPAC-IUB Commission on Biochemical Nomenclature. Carotenoids." Ed. O. Isler. Birkhäuser Verlag, Basel: 851–64.
9. IUPAC. 1975. "IUPAC Commission on the Nomenclature of Organic Chemistry (CNOC) and IUPAC-IUB Commission on Biochemical Nomenclature (CBN), Nomenclature of carotenoids (Rules approved 1974)." *Pure Appl Chem* 41:107–31.
10. Weedon, B. C. L., and Moss G. P. 1995. "Structure and Nomenclature," in: *Carotenoids, Volume 1A: Isolation and Analysis*, pp. 27–70. Eds Britton, G., Liaaen-Jensen, S., and Pfander, H., Birkhäuser Verlag, Basel.
11. Englert, G., Brown, B. O., Moss, G. P., Weedon, B. C. L, Britton, G., Goodwin, T. W., Simpson, K. L., and Williams, R. J. H. 1979. "Prolycopene, a tetra-cis carotene with two hindered cis double bonds." *J Chem Soc Chem Commun* 12:545–7.
12. Britton, G. 1995. "UV/Visible Spectroscopy," in: *Carotenoids, Volume 1B: Spectroscopy*, pp. 13–62. Eds Britton, G., Liaaen-Jensen, S., and Pfander, H., Birkhäuser Verlag, Basel.
13. Hengartner, U., Bernhard, K., Meyer, K., Englert, G., and Glinz, E., 1992. "Synthesis, isolation, and NMR-spectroscopy characterization of fourteen (Z)-isomers of lycopene and of some acetylenic didehydro- and tetradehydrolycopenes." *Helv Chim Acta* 75:1848–65.

14. Goupy, P., Reynaud, E., Dangles, O., and Caris-Veyrat, C. 2011. "Antioxidant activity of (all-E)-lycopene and synthetic apo-lycopenoids in a chemical model of oxidative stress in the gastro-intestinal tract." *New J Chem* 36:575–87.

15. Caris-Veyrat, C., Schmid, A., Carail, M., and Böhm, V. 2003. "Cleavage products of lycopene produced by in vitro oxidations: Characterization and mechanisms of formation." *J Agric Food Chem* 51:7318–25.

16. Chanforan, C., Loonis, M., More, N., Caris-Veyrat, C., and Dufour, C. 2012. "The impact of industrial processing on health-beneficial tomato microconstituents." *Food Chem* 134:1786–95.

17. Kopec, R. E., Riedl, K. M., Harrison, E. H., Curley, R. W., Jr., Hruszkewycz, D. P., Clinton, S. K., and Schwartz, S. J. 2010. "Identification and quantification of apolycopenals in fruits, vegetables, and human plasma." *J Agric Food Chem* 58:3290–6.

18. Carail, M. and Caris-Veyrat, C. 2006. "Carotenoid oxidation products: From villain to saviour?" *Pure App Chem* 78:1493–1503.

19. Khachik, F., Beecher, G. R., and Smith, J. C., Jr. 1995. "Lutein, lycopene, and their oxidative metabolites in chemoprevention of cancer." *J Cell Biochem* 22:236–46.

20. Wang, X. D. 2004. "Carotenoid oxidative/degradative products and their biological activities," in: *Carotenoids in Healths and Disease*, pp. 313–35. Eds Krinsky, N. I., Mayne, S. T., and Sies, H., CRC Press, Boca Raton, FL.

21. Reynaud, E., Aydemir, G., Rühl, R., Dangles, O., and Caris-Veyrat, C. 2011. "Organic synthesis of new putative lycopene metabolites and preliminary investigation of their cell-signalling effects." *J Agric Food Chem* 59:1457–63.

22. Müller, L., Reynaud, E., Goupy, P., Caris-Veyrat, C., and Böhm, V. 2012. "Do apolycopenoids have antioxidant activity in vitro?" *J Am Oil Chem Soc* 88:849–58.

23. Linnewiel, K., Ernst, H., Caris-Veyrat, C., Ben-Dor, A., Kampf, A., Salman, H., Danilenko, M., Levy, J., and Sharoni, Y. 2009. "Structure activity relationship of carotenoid derivatives in activation of the electrophile/antioxidant response element transcription system." *Free Radical Biol Med* 47:659–67.

24. Gouranton, E., Aydemir, G., Reynaud, E., Marcotorchino, J., Malezet, C., Caris-Veyrat, C., Blomhoff, R., Landrier, J. F., and Rühl, R. 2011. "Apo-10'-lycopenoic acid impacts adipose tissue biology via the retinoic acid receptors." *Biochim Biophys Acta* 1811:1105–14.

25. Rao, A. V., Ray, and M. R., Rao, L. G. 2006. "Lycopene." *Adv Food Nutr Res* 51:99–164.

26. Frenich, A. G., Torres, M. E., Vega, A. B., Vidal, J. L., and Bolanos, P. P. 2005. "Determination of ascorbic acid and carotenoids in food commodities by liquid chromatography with mass spectrometry detection." *J Agric Food Chem* 53:7371–6.

27. Lugasi, A., Biro, L., Hovarie, J., Sagi, K.-V., Brandt S., and Barna E. 2003. "Lycopene content of foods and lycopene intake in two groups of the Hungarian population." *Nutr Res* 23:1035–44.

28. Setiawan, B., Sulaeman, A., Giraud, D. W., and Driskell, J. A. "Carotenoid content of selected Indonesian fruits." *J Food Comp Anal* 14:169–76.

29. Arab, L. and Steck, S. 2000. "Lycopene and cardiovascular disease." *Am J Clin Nutr* 71:1691S–1695S.

30. Clinton, S. K. 1998. "Lycopene: Chemistry, biology, and implications for human health and disease." *Nutr Rev* 56:35–51.

31. Lucarini, M., Lanzi, S., D'Evoli, L., Aguzzi, A., and Lombardi-Boccia, G. 2006. "Intake of vitamin A and carotenoids from the Italian population—Results of an Italian total diet study." *Int J Vitam Nutr Res* 76:103–109.

32. Hart, D. J. and Scott, K. J. 1995. "Development and evaluation of an HPLC method for the analysis of carotenoid in foods, and the measurement of the carotenoid content of vegetables and fruits commonly consumed in the UK." *Food Chem* 54:101–11.

33. D'Souza, M. C., Singha, S., and Ingle, M. 1992. "Lycopene concentration of tomato fruit can be estimated from chromaticity values." *HortScience* 27:465–6.
34. Sharma, S. K. and Le Maguer, M. 1996. "Lycopene in tomatoes and tomato pulp fractions." *Ital J Food Sci* 2:107–13.
35. Agarwal, A., Shen, H., Agarwal, S., and Rao, A. V. 2001. "Lycopene content of tomato products: Its stability, bioavailability and in vivo antioxidant properties." *J Med Food* 4:9–15.
36. Brown, E. D., Micozzi, M. S., Craft, N. E., Bieri, J. G., Beecher, G., Edwards, B. K., Rose, A., Taylor, P. R., and Smith, J. C., Jr. 1989. "Plasma carotenoids in normal men after a single ingestion of vegetables or purified beta-carotene." *Am J Clin Nutr* 49:1258–65.
37. Gärtner, C., Stahl, W., and Sies, H. 1997. "Lycopene is more bioavailable from tomato paste than from fresh tomatoes." *Am J Clin Nutr* 66:116–22.
38. Stahl, W. and Sies, H. 1992. "Uptake of lycopene and its geometrical isomers is greater from heat-processed than from unprocessed tomato juice in humans." *J Nutr* 122:2161–6.
39. Boileau, T. W., Boileau, A. C., and Erdman, J. W. Jr. 2002. "Bioavailability of all-trans and cis-isomers of lycopene." *Exp Biol Med (Maywood)* 227:914–19.
40. Porrini, M., Riso, P., and Testolin, G. 1998. "Absorption of lycopene from single or daily portions of raw and processed tomato." *Br J Nutr* 80:353–61.
41. Sanchez-Moreno, C., Cano, M. P., de Ancos, B., Plaza, L., Olmedilla, B., Granado, F., and Martin, A. 2006. "Mediterranean vegetable soup consumption increases plasma vitamin C and decreases F2-isoprostanes, prostaglandin E2 and monocyte chemotactic protein-1 in healthy humans." *J Nutr Biochem* 17:183–9.
42. Kerkhofs, N. S., Lister, C. E., and Savage, G. P. 2005. "Change in colour and antioxidant content of tomato cultivars following forced-air drying." *Plant Foods Hum Nutr* 60:117–21.
43. Kruth, H. S., Huang, W., Ishii, I., and Zhang, W. Y. 2002. "Macrophage foam cell formation with native low density lipoprotein." *J Biol Chem* 277:34573–80.
44. Nofer, J. R., Brodde, M. F., and Kehrel, B. E. 2010. "High-density lipoproteins, platelets and the pathogenesis of atherosclerosis." *Clin Exp Pharmacol Physiol* 37:726–35.
45. Jessup, W., Kritharides, L., and Stocker, R. 2004. "Lipid oxidation in atherogenesis: An overview." *Biochem Soc Trans* 32:134–8.
46. Libby, P., Ridker, P. M., and Maseri, A. 2002. "Inflammation and atherosclerosis." *Circulation* 105:1135–43.
47. Paoletti, R., Gotto, A. M., Jr., and Hajjar, D. P. 2004. "Inflammation in atherosclerosis and implications for therapy." *Circulation* 109:III-20-III-26.
48. Shah, P. K. 1999. "Plaque disruption and thrombosis. Potential role of inflammation and infection." *Cardiol Clin* 17:271–81.
49. Segrest, J. P., Jones, M. K., De Loof, H., and Dashti, N. 2001. "Structure of apolipoprotein B-100 in low density lipoproteins." *J Lipid Res* 42:1346–67.
50. McNamara, J. R., Small, D. M., Li, Z., and Schaefer, E. J. 1996. "Differences in LDL subspecies involve alterations in lipid composition and conformational changes in apolipoprotein." *Brit J Lipid Res* 37:1924–35.
51. Rizzo, M. and Berneis, K. 2007. "Who needs to care about small, dense low-density lipoproteins?" *Int J Clin Pract* 61:1949–56.
52. Müller, L., Caris-Veyrat, C., Lowe, G., and Böhm, V. 2016. "Lycopene and its antioxidant role in the prevention of cardiovascular diseases—A critical review." *Crit Rev Food Sci Nutr* 56:1868–79.
53. Miller, N. J., Sampson, J., Candeias, L. P., Bramley, P. M., and Rice-Evans, C. A. 1996. "Antioxidant activities of carotenes and xanthophylls." *FEBS Lett* 384:240–2.

54. Woodall, A. A., Britton, G., and Jackson, M. J. 1997. "Carotenoids and protection of phospholipids in solution or in liposomes against oxidation by peroxyl radicals: Relationship between carotenoid structure and protective ability." *Biochim Biophys Acta* 1336:575–86.

55. Stahl, W. and Sies, H. 2003. "Antioxidant activity of carotenoids." *Mol Aspects Med* 24:345–51.

56. Kong, K.-W., Khoo, H.-E., Prasad, K. N., Ismail, A., Tan C.-P., and Rajab, N.F. 2010. "Revealing the power of the natural red pigment lycopene." *Molecules* 15:959–87.

57. Krinsky, N. I. and and Yeum, K.-J. 2003. "Carotenoid-radical interactions." *Biochem Biophys Res Commun* 305:754–60.

58. Wagner, K. H. and Elmadfa, I. 2003. "Biological relevance of terpenoids. Overview focusing on mono-, di- and tetraterpenes." *Ann Nutr Metab* 47:95–106.

59. Sundquist, A. R., Briviba, K., and Sies, H. 1994. "Singlet oxygen quenching by carotenoids." *Methods Enzymol* 234:384–8.

60. Yamaguchi, L. F., Martinez, G. R., Catalani, L. H., Medeiros, M. H., and Di Mascio, P. 1999. "Lycopene entrapped in human albumin protects 2'-deoxyguanosine against singlet oxygen damage." *Arch Latinam Nutr* 49:12S–20S.

61. Cantrell, A., McGarvey, D. J., Truscott, T. G., Rancan, F., and Böhm, F. 2003. "Singlet oxygen quenching by dietary carotenoids in a model membrane environment." *Arch Biochem Biophys* 412:47–54.

62. McNulty, H., Jacob, R. F., and Mason, R. P. 2008. "Biologic activity of carotenoids related to distinct membrane physicochemical interactions." *Am J Cardiol* 101:20D–29D.

63. Conn, P. F., Lambert, C., Land, E. J., Schalch, W., and Truscott, T. G. 1992. "Carotene-oxygen radical interactions." *Free Radical Res Commun* 16:401–408.

64. Galano, A., Vargas, R., and Martinez, A. 2010. "Carotenoids can act as antioxidants by oxidizing the superoxide radical anion." *Phys Chem Chem Phys* 12:193–200.

65. Lu, Y., Etoh, H., Watanabe, N., Ina, K., Ukai, N., Oshima, S., Ojima, F., Sakamoto, H., and Ishiguro, Y. 1995. "A new carotenoid, hydrogen peroxide oxidation products from lycopene." *Biosci Biotechnol Biochem* 59:2153–5.

66. Khachik, F., Spangler, C. J., Smith, J. C., Jr., and Canfield, L. M. 1997. "Identification, quantification, and relative concentrations of carotenoids and their metabolites in human milk and serum." *Anal Chem* 69:1873–81.

67. Yeum, K.-J., Aldini, G., Chung, H.-Y., Krinsky, N. I., and Russell, R. M. 2003. "The activities of antioxidant nutrients in human plasma depend on the localization of attacking radical species." *J Nutr* 133:2688–91.

68. Guo, J. J. and Hu, C. H. 2010. "Mechanism of chain termination in lipid peroxidation by carotenes: A theoretical study." *J Phys Chem B* 114:16948–58.

69. Stahl, W., Junghans, A., de Boer, B., Driomina, E. S., Briviba, K., and Sies, H. 1998. "Carotenoid mixtures protect multilamellar liposomes against oxidative damage: Synergistic effects of lycopene and lutein." *FEBS Lett* 427:305–308.

70. Woodall, A. A., Britton, G., and Jackson, M. J. 1995. "Antioxidant activity of carotenoids in phosphatidylcholine vesicles: Chemical and structural considerations." *Biochem Soc Trans* 23:133S.

71. Özyürek, M., Bektasoglu, B., Güçlü, K., Güngör, N., and Apak, R. 2008. "Simultaneous total antioxidant capacity assay of lipophilic and hydrophilic antioxidants in the same acetone-water solution containing 2% methyl-beta-cyclodextrin using the cupric reducing antioxidant capacity (CUPRAC) method." *Anal Chim Acta* 630:28–39.

72. Liu, D., Shi, J., Colina Ibarra, A., Kakuda, Y., and Xue, S. J. 2008. "The scavenging capacity and synergistic effects of lycopene, vitamin E, vitamin C, and β-carotene mixtures on the DPPH free radical." *LWT-Food Sci Technol* 41:1344–4139.

73. Halliwell, B., Zhao, K., and Whiteman, M. 2000. "The gastrointestinal tract: A major site of antioxidant action?" *Free Radical Res* 33:819–30.

74. Benzie, I. F. and Strain, J. J. 1996. "The ferric reducing ability of plasma (FRAP) as a measure of 'antioxidant power': The FRAP assay." *Anal Biochem* 239:70–6.

75. Pulido, R., Bravo, L., and Saura-Calixto, F. 2000. "Antioxidant activity of dietary polyphenols as determined by a modified ferric reducing/antioxidant power assay." *J Agric Food Chem* 48:3396–402.

76. Müller, L. and Böhm, V. 2011. "Antioxidant activity of β-carotene compounds in different in vitro assays." *Molecules* 16:1055–69.

77. Tsimikas, S. and Miller, Y. I. 2011. "Oxidative modification of lipoproteins: Mechanisms, role in inflammation and potential clinical applications in cardiovascular disease." *Curr Pharm Design* 17:27–37.

78. Hazen, S. L. and Heinecke, J. W. 1997. "3-Chlorotyrosine, a specific marker of myelo peroxidase-catalyzed oxidation, is markedly elevated in low density lipoprotein isolated from human atherosclerotic intima." *J Clin Invest* 99:2075–81.

79. Winterbourn, C. C. and Kettle, A. J. 2000. "Biomarkers of myeloperoxidase-derived hypochlorous acid." *Free Radical Biol Med* 29:403–409.

80. Zhang, R., Brennan, M. L., Shen, Z., MacPherson, J. C., Schmitt, D., Molenda, C. E., and Hazen S. L. 2002. "Myeloperoxidase functions as a major enzymatic catalyst for initiation of lipid peroxidation at sites of inflammation." *J Biol Chem* 277:46116–22.

81. Candeias, L. P., Patel, K. B., Stratford, M. R., and Wardman. P. 1993. "Free hydroxyl radicals are formed on reaction between the neutrophil-derived species superoxide anion and hypochlorous acid." *FEBS Lett* 333:151–3.

82. Pennathur, S., Maitra, D., Byun, J., Sliskovic, I., Abdulhamid, I., Saed, G. M., Diamond, M. P., and Abu-Soud, H. M. 2010. "Potent antioxidative activity of lycopene: A potential role in scavenging hypochlorous acid." *Free Radical Biol Med* 49:205–13.

83. Mortensen, A., Skibsted, L. H., Sampson, J., Rice-Evans, C., and Everett, S. A. 1997. "Comparative mechanisms and rates of free radical scavenging by carotenoid antioxidants." *FEBS Lett* 418:91–7.

84. Rodriguez-Amaya, D. B. 2001. *A guide to carotenoid analysis in food*, ILSI Press, Washington.

85. Shi, J., and Le Maguer, M. 2000. "Lycopene in tomatoes: Chemical and physical properties affected by food processing." *Crit Rev Biotechnol* 20:293–334.

86. Henry, L. K., Catignani, G. L., and Schwartz, S. J. 1998. "Oxidative degradation kinetics of lycopene, lutein, and 9-cis and all-trans β-carotene." *J Am Oil Chem Soc* 75:823–9.

87. Boon, C. S., McClements, D. J., Weiss, J., and Decker, E. A. 2009. "Role of iron and hydroperoxides in the degradation of lycopene in oil-in-water emulsions." *J Agric Food Chem* 57:2993–8.

88. Schierle, J., Bretzel, W., Bühler, I., Faccin, N., Hess, D., Steiner, K., and Schüep, W. 1997. "Content and isomeric ratio of lycopene in food and human blood plasma." *Food Chem* 59:459–65.

89. Müller, L., Goupy, P., Fröhlich, K., Dangles, O., Caris-Veyrat, C., and Böhm, V. 2011. "Comparative study on antioxidant activity of lycopene (Z)-isomers in different assays." *J Agric Food Chem* 59:4504–11.

90. Fröhlich, K., Kaufmann, K., Bitsch, R., and Böhm, V. 2006. "Effects of ingestion of tomatoes, tomato juice and tomato puree on contents of lycopene isomers, tocopherols and ascorbic acid in human plasma as well as on lycopene isomer pattern." *Brit J Nutr* 95:734–41.

91. Palozza, P., Simone, R., Catalano, A., Boninsegna, A., Böhm, V., Fröhlich, K., Mele, M. C., Monego G, and Ranelletti, F. O. 2010. "Lycopene prevents 7-ketocholesterol-induced oxidative stress, cell cycle arrest and apoptosis in human macrophages." *J Nutr Biochem* 21:34–46.

92. Tang, X., Yang, X., Peng, Y., and Lin, J. 2009. "Protective effects of lycopene against H₂O₂-induced oxidative injury and apoptosis in human endothelial cells." *Cardiovasc Drugs Ther* 23:439–48.
93. Hung, C. F., Huang, T. F., Chen, B. H., Shieh, J. M., Wu, P. H., and Wu, W. B. 2008. "Lycopene inhibits TNF-alpha-induced endothelial ICAM-1 expression and monocyte-endothelial adhesion." *Eur J Pharmacol* 586:275–82.
94. Yamashita, S. and Matsuzawa, Y. 2009. "Where are we with probucol: A new life for an old drug?" *Atherosclerosis* 207:16–23.
95. Palozza, P., Parrone, N., Simone, R. E., and Catalano, A. 2010. "Lycopene in atherosclerosis prevention: An integrated scheme of the potential mechanisms of action from cell culture studies." *Arch Biochem Biophys* 504:26–33.
96. Palozza, P., Simone, R., Catalano, A., Parrone, N., Monego, G., and Ranelletti, F. O. 2011. "Lycopene regulation of cholesterol synthesis and efflux in human macrophages." *J Nutr Biochem* 22:971–8.
97. Lo, H.-M., Hung, C.-F., Tseng, Y.-L., Chen, B.-H., Jian, J.-S., and Wu, W. B. 2007. "Lycopene binds PDGF-BB and inhibits PDGF-BB-induced intracellular signaling transduction pathway in rat smooth muscle cells." *Biochem Pharmacol* 74:54–63.
98. Hsiao, G., Wang, Y., Tzu, N. H., Fong, T. H., Shen, M. Y., Lin, K. H., Chou, D. S., and Sheu, J. R. 2005. "Inhibitory effects of lycopene on in vitro platelet activation and in vivo prevention of thrombus formation." *J Labatory Clin Med* 146:216–26.
99. Thies, F., Masson, L. F., Rudd, A., Vaughan, N., Tsang, C., Brittenden, J., Simpson, W. G., Duthie, S., Horgan, G. W., and Duthie, G. 2012. "Effect of a tomato-rich diet on markers of cardiovascular disease risk in moderately overweight, disease-free, middle-aged adults: A randomized controlled trial." *Am J Clin Nutr* 95:1013–22.
100. McEneny, J., Wade, L., Young, I. S., Masson, L., Duthie, G., McGinty, A., McMaster, C., and Thies, F. 2013. "Lycopene intervention reduces inflammation and improves HDL functionality in moderately overweight middle-aged individuals." *J Nutr Biochem* 24:163–8.
101. Martínez-Tómas, R., Pérez-Llamas, F., Sánchez-Campillo, M., Gonzaléz-Silvera, D., Cascales, A. I., García-Fernández, M., López-Jiménez, J. A., Zamora Navarro, S., Burgos, M. I., López-Azorín, F., Wellner, A., Avilés Plaza, F., Bialek, L., Aminger, M., and Larqué, E. 2012. "Daily intake of fruit and vegetable soups processed in different ways increases human serum β-carotene and lycopene concentrations and reduces levels of several oxidative stress markers in healthy subjects." *Food Chem* 134:127–33.
102. Abete, I., Perez-Cornago, A., Navas-Carretero, S., Bondia-Pons, I., Zulet, M. A., and Martinez, J. A. 2013. "A regular lycopene enriched tomato sauce consumption influences antioxidant status of healthy young-subjects: A crossover study." *J Funct Foods* 5:28–35.
103. Gajendragadkar, P. R., Hubsch, A., Mäki-Petäjä, K. M. Serg, M., Wilkinson, I. B., and Cheriyan, J. 2014. "Effects of oral lycopene supplementation on vascular function in patients with cardiovascular disease and healthy volunteers: A randomised controlled trial." *PLoS ONE* 6:e99070.
104. Ghavipour, M., Sotoudeh, G., and Ghorbani, M. 2015. "Tomato juice consumption improves blood antioxidative biomarkers in overweight and obese females." *Clin Nutr* 34:805–809.
105. Burton-Freeman, B., Edirisinghe, I., Cappozzo, J., Banaszewski, K., Giordano, R., Kappagoda, C. T., Cao, Y., and Kris-Etherton, P. 2015. "Processed tomato products and risk factors for cardiovascular disease." *Nutr Aging* 3:193–201.
106. Hirose, A., Terauchi, M., Tamura, M., Akiyoshi, M., Owa, Y., Kato, K., and Kubota, T. 2015. "Tomato juice intake increases resting energy expenditure and improves hypertriglyceridemia in middle-aged women: An open-label, single-arm study." *Nutr J* 14:34.

5 Modulation of Inflammation by Tomatoes and Lycopene in the Context of Cardiometabolic Diseases

Jean-François Landrier, Franck Tourniaire,
Soumia Fenni, Charles Desmarchelier,
and Patrick Borel

CONTENTS

5.1 INTRODUCTION

Lycopene is a lipophilic pigment that is responsible for the red color of various fruits, such as tomato, water melon, guava, or grapefruit. It belongs to the carotenoid family, and more specifically to the carotene class, but it is not a precursor of retinol (vitamin A). In Western diets, tomatoes and tomato products represent the major sources of lycopene (Rao, Ray, and Rao 2006).

Whereas all-trans lycopene is the main dietary form (Richelle et al. 2012), lycopene can be found in human blood and tissues under the all-*trans* form and as *cis*-isomers and the conformation of the molecule could modulate its activity.

It is unclear whether the change in its conformation results from chemical and/ or enzymatic processes but it has been observed that the percentage of *cis* isoforms increased following the absorption of all-*trans* lycopene (Richelle et al. 2010; Moran et al. 2015) and represented around a third of the total lycopene (Ross et al. 2011). The ratio between *trans* and *cis* isomers seems to vary depending on the tissues investigated: in humans, it is close to 1 in plasma and 2 in the human liver (Khachik et al. 2002).

Beside native lycopene, several metabolites have been described in various tissues and biological fluids. Indeed, Khachik et al. have identified a metabolite of lycopene named 2,6-cyclolycopene-1,5 diol that is present in human plasma, liver, and milk (Khachik et al. 1998; Khachik et al. 2002). Studies have identified circulating and tissue lycopene metabolites (aldehydes) produced by the shortening of the isoprenoid chain of the molecule: apo-6′, apo-8′, and apo-12′-lycopenals have been found in the liver of rodents fed with lycopene (Gajic et al. 2006) or tomato products (Tan et al. 2014; Martin-Pozuelo et al. 2015). These compounds have also been identified in human plasma following the consumption of tomato juice, together with apo-10′ and apo-14′-lycopenals (Kopec et al. 2010). However, it was not possible to find out whether they were produced endogenously from lycopene since they were also present in the tomato products.

The identification of the genetic sequences of two isoforms of carotenoid oxygenases (beta-carotene oxygenase 1 and 2, BCO1 and BCO2) in mammals has allowed the synthesis of the corresponding recombinant proteins. Studies have been undertaken to characterize their ability to cleave lycopene *in vitro*. BCO1 activity is located in the cytoplasm and is in charge of the first step of synthesis of retinoids (i.e., retinal) from provitamin A carotenoids (Redmond et al. 2001). Redmond et al. have reported a weak activity of recombinant mouse BCO1 towards lycopene (Redmond et al. 2001) but more recently, dela Sena et al. found that recombinant human BCO1 is able to cleave lycopene at position 15,15′ almost as efficiently as beta-carotene (dela Sena et al. 2013). BCO2 activity is restricted to the mitochondrial membrane, where it prevents non-provitamin A carotenoid accumulation, which can impair the respiratory chain (Amengual et al. 2011; Palczewski et al. 2014; Raghuvanshi et al. 2015). BCO2 seems to play an important role in lycopene metabolism since $Bco2^{-/-}$ mice fed with tomato/lycopene-enriched diets accumulate more lycopene in their liver than their wild-type or $Bco1^{-/-}$ littermates (Ford et al. 2010; Ford, Elsen, and Erdman 2013; Tan et al. 2014), suggesting a higher absorption and/or a decreased degradation of lycopene. BCO2 affinity seems to vary across species: recombinant ferret BCO2 displayed activity only towards 5-*cis* and 13-*cis* lycopene isoforms, but not towards the all-*trans* isoform (Hu et al. 2006). Previous work from Kiefer et al. indicated the production of apolycopenal in a lycopene-producing E. coli strain expressing mouse BCO2 (Kiefer et al. 2001). By analogy with retinal (that can be further transformed to either retinol or retinoic acid), it has been suggested that apolycopenals could be metabolized to the corresponding acid or alcohol forms (i.e., apolycopenoic acid or apolycopenol). Hu et al. have been able to produce apo-10′-lycopenol and apo-10′-lycopenoic acid by incubating apo-10′-lycopenal with ferret liver homogenates, suggesting that the enzymatic equipment allowing such metabolisation is present in mammals (Hu et al. 2006). However, the *in vivo* relevance of apocarotenoic acids/apocarotenols remains questionable since, so far, they have only been observed in $Bco1^{-/-}$ mice fed with xanthophyll or beta-carotene-rich diets (Amengual et al. 2013).

In humans, adipose tissue and liver are the main lycopene storage tissues, and contain about 60% and 30%, respectively, of the total lycopene body stores (Chung et al. 2009; Landrier, Marcotorchino, and Tourniaire 2012; Moran, Erdman, and Clinton 2013). Furthermore, it has been observed that lycopene bioavailability and deposition in the liver of rodents is increased in the context of a high-fat diet (Bernal et al. 2013; Martin-Pozuelo et al. 2015). In the liver, the uptake of lycopene is probably mediated via chylomicrons-remnant receptors, i.e., the LDL-receptor (LDLR), the LDL-receptor-related protein 1 (LRP1), and the heparan sulfate proteoglycans (HSPGs) (Dallinga-Thie et al. 2010). In adipose tissue and adipocytes, the uptake of lycopene is mediated, at least in part, by CD36 (Moussa et al. 2011) and is not related to its physicochemical properties (Sy et al. 2012). Lycopene is then stored in the adipocyte lipid droplets and in membranes (Gouranton et al. 2008). It is generally assumed that lycopene in these tissues could mediate some biological effects. It is worth noting that these tissues are also key sites in the development of inflammatory processes in the context of the metabolic syndrome and more generally for cardiometabolic disorders.

In terms of biological function, lycopene is well known for its antioxidant properties (Palozza et al. 2010). It displays anti-free radical properties mediated by 11 conjugated double bonds. It is essential in scavenging lipid peroxyl radicals, reactive oxygen species, and nitric oxide (Engelmann, Clinton, and Erdman 2011). In addition, several studies have shown that lycopene displays anti-inflammatory effects in several tissues and in several pathophysiological disorders, which are highly relevant in the context of cardiometabolic disorders. This is notably the case in adipocytes and adipose tissue (Gouranton, Thabuis et al. 2011; Marcotorchino et al. 2012; Luvizotto Rde et al. 2013; Singh et al. 2016; Fenni et al. 2017), in the liver (Wang 2012; Ip et al. 2014), and in arterial wall cells (Armoza et al. 2013; Hung et al. 2008).

Among others, these anti-inflammatory effects could be responsible for the numerous health effects attributed to lycopene (Wang 2012; Story et al. 2010) in the field of liver steatosis (Wang 2012), cardiovascular diseases (Muller et al. 2015; Thies et al. 2012), adiposity and obesity (Bonet et al. 2015; Landrier, Marcotorchino, and Tourniaire 2012). Indeed, inflammation appears as a major hallmark and is intimately related to all cardiometabolic disorders, including type-2 diabetes, cardiovascular diseases, and metabolic syndrome (Esser, Paquot, and Scheen 2015). It is noteworthy that obesity is a major driving force of inflammation and thus constitutes a major risk factor for the development of these disorders. Obesity is characterized by a chronic low-grade or metabolic inflammation, which mainly originates from adipose tissue enlargement (Gregor and Hotamisligil 2011). Pro-inflammatory cytokines, such as tumor necrosis factor alpha (TNF-α), interleukin (IL)-6, and IL-1β, and chemokines, such as MCP1/CCL2 or RANTES/CCL5, are produced by adipose tissue (Tourniaire et al. 2013) and are widely acknowledged as important pathophysiological factors implicated in insulin resistance. In addition, these markers of inflammation also stimulate the production of acute phase proteins, such as C-reactive protein (CRP), serum amyloid A, and adhesion molecules, such as soluble intercellular adhesion molecule type 1 (sICAM1) or VCAM, in the liver, which are markers of increased CVD risk (Pearson et al. 2003; Ridker, and Morrow 2003).

The overall objective of the present review is to summarize the knowledge related to the anti-inflammatory effects of lycopene, its metabolites, tomatoes, tomato products, or extracts, from cell culture to human clinical studies that could beneficially impact prevalence of cardiometabolic disorders.

5.2   *IN VITRO* STUDIES

We have demonstrated the ability of lycopene to inhibit the expression of pro-inflammatory cytokines and chemokines *in vitro* in murine and human adipocytes (Gouranton et al. 2011). These data were also reproduced *ex vivo* in adipose tissue explants from mice subjected to a high-fat diet (characterized by low-grade inflammation), and confirmed by another group in human adipocytes (Warnke et al. 2016). The molecular mechanism was investigated and the involvement of NF-κB was confirmed (Gouranton et al. 2011). Similar results (i.e., inhibition of cytokine and chemokine expression in various *in vitro* and *ex vivo* models) were obtained with apo-10′-lycopenoic acid, a metabolite of lycopene (Gouranton, Aydemir et al. 2011). Finally, lycopene attenuated LPS-mediated induction of TNFα in macrophages via NF-κB and JNK, as well as macrophage migration *in vitro* (Marcotorchino et al. 2012). Consequently, lycopene decreased macrophage-induced cytokine, acute phase protein, and chemokine mRNA levels in adipocytes. Similarly, tomato extract suppressed inflammation and pro-inflammatory cytokine production during interaction between adipocytes and macrophages (Kim et al. 2015).

In macrophages, lycopene (Zou et al. 2013; Hadad and Levy 2012; Feng, Ling, and Duan 2010) and tomato aqueous extracts (Schwager et al. 2016; Navarrete, Alarcon, and Palomo 2015) suppressed the pro-inflammatory response (including cytokines and interleukins such as TNFα, IL-1β, IL-6, etc.) in RAW264.7 cells incubated with LPS, probably via an inhibition of the NF-κB pathway (Hadad and Levy 2012). In THP-1 cells, lycopene also prevented the activation of the pro-inflammatory cascade mediated by oxysterols via an inhibition of NF-κB and an increase in PPARγ expression. Lycopene synthetic metabolites apo-10′-lycopenoic acid and apo-14′-lycopenoic acid repressed NF-κB activation induced by cigarette smoke extract in THP-1 macrophage cells (Catalano et al. 2013) and MMP9 expression (Palozza et al. 2012).

The effects of lycopene and tomato products have been studied in several vascular endothelial cell models, including HUVEC cells. In these cells, tomato oleoresin and purified lycopene significantly inhibited the TNFα-mediated expression of several adhesion molecules, including ICAM-1 and VCAM-1, via a decrease in NF-κB signaling (decrease in IκB protein concentration and p65 phosphorylation level) (Armoza et al. 2013; Hung et al. 2008). Tomato extracts also limited the gene expression of pro-inflammatory cytokines and chemokines (TNFα and IL-8) and induced the gene expression of anti-inflammatory cytokines, such as IL-10, after stimulation by TNFα, possibly by targeting NF-κB signaling (Hazewindus et al. 2014; Di Tomo et al. 2012). This effect was associated with a limitation of monocyte migration. Similarly, tomato aqueous extracts blunted the production of a large range of interleukins and chemokines (Schwager et al. 2016). Interestingly, lycopene has also been depicted as able to attenuate the activation of cell signaling related to inflammation, such as ERK, p38 (Chen et al. 2012), PI3K/AKT, or NRF2 (Sung et al. 2015). Other pro-inflammatory stimuli have been tested regarding the ability of lycopene to blunt the pro-inflammatory responses to HMGB1 (high-mobility group box 1; (Lee et al. 2012)) or LPS (Wang et al. 2013).

5.3 PRECLINICAL STUDIES

5.3.1 CARDIOVASCULAR DISEASES

In the context of hyperhomocysteinemia, a risk factor for atherosclerosis, lycopene supplementation (10–20 mg/kg of body weight p.o. for ten weeks) has been shown to reduce aortic lipid depots and decrease the level of numerous circulating inflammatory markers (serum VCAM1, MCP1, IL-8) (Liu et al. 2007).

In vivo myocardial ischemia/reperfusion (I/R) injury models are used to study the potential of compounds to limit damages associated with this event. Whereas several studies have been published regarding the benefits of lycopene treatment on I/R induced cardiac dysfunctions (Xu et al. 2015), only a few have documented the consequences on inflammatory aspects. It was shown in rats by Bansal et al. that oral administration of lycopene for 31 days (1 mg/kg), before inducing a chirurgical occlusion of the left anterior ascending coronary artery followed by a reperfusion, was able to diminish damages associated with I/R, as well as inflammatory cell infiltration in myocardial tissue (Bansal et al. 2006). He et al. have demonstrated that chronic lycopene intake (10 mg/kg body weight daily for four weeks) can protect against inflammation associated with post-myocardial infarction remodeling, as shown by diminished expression of Tnf and Il1b mRNAs and decreased NF-κB activation in mice (He et al. 2015). Using an opposite approach, i.e. inducing cardiac remodeling first then administering lycopene (40 mg/kg body weight daily for 28 days), Wang et al. found that lycopene could partially maintain cardiac function and limit p38 activation pathway (Wang et al. 2014). This was in agreement with what was previously reported by another group using another model of myocardial infarction (induced by isoproterenol); lycopene intake also diminished the infarction area size and inflammation assessed by measurement of circulating CRP level and tissular myeloperoxidase activity (Upaganlawar, Gandhi, and Balaraman 2010). In a recent study, Tong et al. showed that acute (intravenous) lycopene administration just after ischemia diminished the size of the infarct and that it was associated with a decrease in the activation of two inflammatory signaling pathways, JNK and p42/44 MAPK (Tong et al. 2016).

5.3.2 OBESITY MODELS/METABOLIC SYNDROME/ NON-ALCOHOLIC LIVER STEATOSIS

Several groups have explored the anti-inflammatory effects of tomato extract and/or lycopene intake—mainly in rats—in the context of obesity. In these studies, obesity was induced using high-fat diets, associated or not with administration of lycopene, and the consequences of lycopene intake on obesity onset was observed. Bahcecioglu and colleagues have used two doses of lycopene (two and four mg/kg bw/d for six weeks) and reported beneficial effects of lycopene administration on insulin sensitivity, liver steatosis, and inflammation in rats only at a dose of 2 mg/kg/d (Bahcecioglu et al. 2010). However, the highest dose led to fewer beneficial changes, suggesting that there might be an optimal dosage of lycopene in this context. A similar observation was made in a study involving mice fed with a high-fat and high-cholesterol (HF/HChol) diet supplemented with two doses of dried tomato peel (9% or 17%) for 12 weeks (Zidani et al. 2017).

Overall, all studies have highlighted the decrease in local and/or systemic inflammation following tomato product/lycopene consumption, except in the studies by the group of Periago (Bernal et al. 2013; Martin-Pozuelo et al. 2015). However, it should be noted that in these papers, the HF/HChol diet had no effect on circulating TNFα or IL-6 compared with control diet, neither did it induce signs of inflammation in the liver, as assessed histologically. Conversely, a similar study performed with lycopene (isolated from algae or tomato) supplementation of a HF/HChol diet led to a reduced degree of liver steatosis and atherosclerotic depot in the aorta, as well as anti-inflammatory effects (decreased serum levels of ceruloplasmine, CRP, myeloperoxidase, and decreased 15-lipoxygenase and cyclo-oxygenase activities in PBMCs) (Renju, Kurup, and Saritha Kumari 2014). Interestingly, the lycopene from algae origin displayed more beneficial effects than the one isolated from tomato, suggesting that synergetic effects with other compounds present in these matrices play a role in this effect.

Others have used a model of diet-induced obesity (DIO) involving a high-fat diet plus sucrose, supplemented or not with lycopene, in rats. Under these conditions, it was observed that lycopene treatment improved insulin resistance-related parameters (plasma insulin, glycemia) and could limit inflammation at the central level (decreased plasma concentrations of leptin, resistin, and IL6 but not TNFα), in the adipose tissue (decreased mRNA expression of leptin, resistin, Mcp1, and Il6 but not Tnf) (Luvizotto Rde et al. 2013) and also in the brain (Yin et al. 2014). Pierine et al. investigated the effect of lycopene supplementation in rats already obese, and again the absence of effect on the circulating level of TNFα was observed, whereas a decreased production was found in the kidney (Pierine et al. 2014).

In a very extensive study, the anti-inflammatory effects of lycopene at a dose of 10 mg/kg of diet limited adiposity, prevented hyperinsulinemia and insulin resistance during the development of DIO in mice. It was found that lycopene limited liver steatosis and inflammation (by decreasing IL6, TNFα, NF-κB, and TLR4) (Singh et al. 2016). Furthermore, it was found that some circulating inflammatory markers (TNFα, IL-1β) were also diminished by lycopene administration, suggesting a generic anti-inflammatory effect. This study also indicated for the first time that lycopene could act on the gut microbiota, as reflected by caecal bacterial abundance and short-chain fatty acid production. The limitation of bacteria and the preservation of gut and colon epithelial integrity associated with obesity and insulin-resistance, could participate in the overall anti-inflammatory properties of lycopene at the systemic level.

In agreement, we also recently showed that supplementation with lycopene or tomato powder (TP) for 12 weeks diminished the hepatic steatosis induced by a high-fat diet (Fenni et al. 2017), and reported an anti-inflammatory effect of lyco-pene or TP in the liver but also an impact on lipid metabolism. We also investigated the impact of lycopene and TP supplementation on adipose tissue inflammatory sta-tus, and reported an overall down-regulation of genes encoding pro-inflammatory markers such as cytokines (Il6, Tnfα), adipokines (resistin, visfatin, leptin), acute phase proteins (Saa3, haptoglobin), and chemokines (Ccl5, Mcp1, Cxcl10), a down-regulation of genes encoding metalloproteinases (Mmp3 and Mmp9), and an up-regulation of genes encoding anti-inflammatory proteins such as Il-10 and Tgf-β. In addition, we confirmed at the protein level that some of these pro-inflammatory markers (IL-6, TNFα, MCP1, CCL5) were reduced compared to HFD conditions. We also observed that lycopene and TP strongly reduced the phosphorylation levels of p65 and IκB *in vivo*, consistently with *in vitro* data, which suggests that the anti-inflammatory effect of lycopene and TP on adipose tissue results from their ability to inhibit NF-κB signaling in adipose tissue. Despite this clear impact of lycopene or TP on the NF-κB signaling pathway, we cannot exclude that this effect was strictly due to the reduced adiposity, which could result in a deactivation of this signaling pathway.

Since apo-10′-lycopenoids are proposed to be the major products of lycopene metabolism via BCO2, the effect of apo-10′-lycopenoic acid towards liver steatosis was investigated in the genetically obese mouse model Ob/Ob (Chung et al. 2012). It was found that apo-10′-lycopenoic acid supplementation diminished the severity of liver steatosis and that this result was mediated by SIRT1, which, among other functions, is able to inhibit NF-κB inflammatory signaling (Schug and Li 2011). In a study comparing the potential of lycopene vs. apo-10′-lycopenoic acid to limit liver steatosis in DIO Bco2$^{-/-}$ mice, Ip et al. identified that these two molecules had distinct modes of action, with apo-10′ lycopenoic acid acting on SIRT1 activity and expression in the liver while lycopene was acting at the level of mesenteric fat through actions of PPARα and PPARγ (Ip et al. 2015). Another nuclear receptor that could mediate anti-inflammatory effects of lycopene and apo-10′-lycopenoic acid is RAR (retinoic acid receptor). We showed that apo-10′-lycopenoic acid was able to transactivate RAR, concomitantly to the prevention of the production of inflamma-tory markers in adipocytes (Gouranton, Aydemir et al. 2011). Furthermore, Aydemir et al. have shown that lycopene administration to mice could lead to the induction of RARE-mediated cell signaling and could even restore vitamin A deficiency (Aydemir et al. 2012).

5.4 HUMAN STUDIES

5.4.1 Cross-Sectional and Prospective Studies

Adjusted circulating concentrations of lycopene as well as many other carot-enoids were inversely associated with CRP concentrations in large studies, such as the National Health and Nutrition Examination Survey III (NHANES III) (Ford et al. 2003; Kritchevsky et al. 2000), and in other studies (Kim et al. 2010). High plasma lycopene concentration (reflected by the highest tertile) was associated with lower plasma CRP, IL-6, and MMP9 concentrations in a general population of 285

Swedish men and women before adjustment, but these associations were lost after adjustments for age, sex, alcohol intake, BMI, systolic blood pressure, and total cholesterol (Ryden et al. 2012). Similarly, lycopene status was inversely associated with IL-6 (Walston et al. 2006) and other markers of endothelial function, such as s-ICAM 1 (van Herpen-Broekmans et al. 2004). In the YALTA prospective study, the initial plasma level of lycopene was inversely associated with sICAM1 protein level 15 years later (Hozawa et al. 2007). Similarly, plasma lycopene concentration was inversely associated with plasma CRP and E-selectin concentrations in men and women of an Aboriginal population and the association remained significant after adjustment for classical confounding factors (Rowley et al. 2003). No relationship was found between plasma lycopene and plasma CRP concentrations in a population of middle-aged and older women (Wang et al. 2008).

5.4.2 Interventions Studies

5.4.2.1 Lycopene Supplementation

Short-term lycopene supplementation (one week; 80 mg/d) in young healthy partici-pants did not modify post-prandial inflammatory status (CRP, ICAM, or VCAM) (Denniss et al. 2008). In obese people, four weeks of lycopene supplementation (Lyc-O-Mato, 30 mg/d) did not modify plasma concentrations of inflammatory markers (TNFα, IL6, CRP) (Markovits, Ben Amotz, and Levy 2009). Similarly, no impact of one-week lycopene supplementation (7 mg/d) was observed in patients with prehy-pertension (Petyaev et al. 2012) and no impact of two weeks' supplementation (lyc-O-Mato 3 capsules/d; approximately 14.64 mg of lycopene/d) was observed on TNFα in a smoker (n = 12) and non-smoker (n = 15) population (Briviba et al. 2004). Seven mg per day of lycopene for two months also did not modify inflammatory markers in CVD patients and healthy volunteers (n = 36 for each group) (Gajendragadkar et al. 2014). In a large study involving 224 healthy middle-aged volunteers supplemented for 12 weeks with lycopene capsules (10 mg/d) or tomato-rich diet, no modification of inflammatory markers was observed (Thies et al. 2012).

On the opposite, it was reported that 12 weeks of a lycopene-rich diet (224–350 mg lycopene/wk) in overweight middle-aged individuals (n = 54) led to a reduction in SAA concentration in the HDL_3 fraction while 12 weeks of lycopene supplementa-tion (70 mg/wk) in the same study group led to a reduction in SAA concentration in both plasma and HDL_3 fraction (McEneny et al. 2013). Lycopene supplementation (7 mg or 15 mg/d for eight weeks) also reduced CRP, ICAM, and VCAM plasma concentrations in healthy men (n = 126) (Kim et al. 2011). In 26 healthy volun-teers, Lyco-O-Mato (5.7 mg of lycopene) for 26 days reduced TNFα production from whole blood after a LPS challenge *in vitro* (Riso et al. 2006).

5.4.2.2 Tomato Supplementation

In type-2 diabetes patients (n = 15), tomato juice supplementation for four weeks (300 ml/d) did not improve CRP, ICAM, or VCAM plasma concentrations compared to placebo (Upritchard, Sutherland, and Mann 2000). It has been established that a tomato-rich diet (300 g/d for one month) had no impact on plasma concentrations of E-selectin, ICAM, and CRP in 103 healthy volunteers (Blum et al. 2007). Tomato juice consumption for two weeks in healthy volunteers (daily dose of lycopene: 20.6 mg) reduced plasma CRP concentration (Jacob et al. 2008). Interestingly, in heathy overweight volunteers, a combination of dietary anti-inflammatory molecules, including tomato extract, resulted in an increase in plasma adiponectin concentration, that displays anti-inflammatory activity, whereas no other modification was observed (Bakker et al. 2010). This result suggests that this anti-inflammatory mix could have a beneficial impact on adipose tissue inflammatory status as reflected by adiponectin (Ruhl and Landrier 2015). Consumption of tomato products blunted the post-prandial induction of plasma IL-6 in healthy women and men (n = 25) (Burton-Freeman et al. 2012). Plasma concentration of lycopene, related to the consumption of a diet rich in fruit and vegetables, was inversely associated with a decrease in the LPS-mediated production of IL-6 in PMBC of overweight women (Yeon, Kim, and Sung 2012). An intake of 330 ml/d of tomato juice for 20 days also significantly reduced plasma concentrations of IL-8 and TNFα in overweight participants while only IL-6 was diminished in obese participants (Ghavipour et al. 2012). A similar reduction in plasma concentrations of plasma IL-6 and TNFα was observed in patients with metabolic syndrome who consumed tomato juice four times a week for a period of two months (Tsitsimpikou et al. 2014). A reduction in CRP concentrations was also observed in women with heart failure but not in men following consumption of a vegetable juice (340 ml/d?) for 30 days (Biddle et al. 2014). In women (n = 30), tomato juice consumption (280 ml/d containing 32.5 mg of lycopene) for eight weeks reduced MCP1 and increased adiponectin (Li et al. 2015), but no control group was used in this study. A similar reduction in MCP1 was observed in healthy people (n = 12) consuming a vegetable soup constituted mainly of tomato (500 ml/d for 14 days) (Sanchez-Moreno et al. 2006). The post-prandial impact of a tomato-rich diet (three types: raw tomatoes, tomato sauce, or tomato sauce with olive oil) was evaluated in healthy volunteers (n = 40): such regimens led to a reduction in MCP1 and to an increase in IL-10 for the three diets, and an increase in IL18 for tomato sauce. IL6 and VCAM were reduced with tomato sauce plus olive oil (Valderas-Martinez et al. 2016). A similar post-prandial impact of carotenoid-rich tomato extract has been depicted in 146 healthy normal-weight individuals regarding the level of oxidized LDL (Deplanque, Muscente-Paque, and Chappuis 2016). A recent meta-analysis reported a significant decrease in IL-6 (standardized mean difference −0.25; p = 0.03) following tomato supplementation (Cheng et al. 2017).

Negative results associated with tomato juice supplementation (330 ml/d for two weeks in 22 healthy men) were reported in only one study, where an increase in TNFα production was observed compared to wash-out periods (Watzl et al. 2003). The origin of this contrasted results is presently unknown.

5.5 CONCLUSION AND PERSPECTIVES

Based on the reported studies, the anti-inflammatory effect of lycopene and/or tomato products is rather clear, at least in preclinical and *in vitro* studies. Indeed, in several cell models related to cardiometabolic health, lycopene or tomato products exert a strong and reproducible inhibition on the expression and secretion of cytokines and interleukins, as well as an impact on several signaling pathways that could participate to the overall effect, including NF-κB. Preclinical studies also strongly support these data. In most cases, an improvement of the inflammatory status has been reported *in vivo*, in key tissues (mainly in the liver and in adipose tissue) and in some cases at the systemic level. All data generated in the context of obesity and its comorbidities suggest that lycopene, and in some cases tomato products, have a beneficial impact on inflammation in key organs and are also associated with clear improvement in metabolic disorders.

Interestingly, this anti-inflammatory effect of lycopene and/or tomato products has been confirmed in some clinical studies at the systemic level. If the effect of isolated lycopene on inflammatory status is still not clear, it is noteworthy that most studies have been realized in healthy people with no inflammation disorders. When tomato supplements are administered to volunteers with an elevated inflammatory status (overweight/obesity or during post-prandial inflammation), it becomes clearer that such a nutritional approach may have beneficial effects on the control of inflammation. In line with this, it is important to specify that lycopene is classically associated with the health benefits of tomato consumption (Canene-Adams et al. 2005). However, these data show that the overlap between lycopene supplementation and tomato product consumption is not perfect regarding their health effects (Burton-Freeman and Sesso 2014).

Several important questions remain unsolved. It is presently not clear if the all-*trans* or *cis* isomers of lycopene display similar activities regarding their anti-inflammatory effect. It is also not clear if lycopene or its metabolites are responsible for the reported effects. Indeed, the activity of lycopene metabolites towards inflammation has so far been barely investigated. It is also noteworthy that in the case of tomato product consumption, it cannot be excluded that the reported effects are only due to lycopene and not to other molecules present, or even to the association of lycopene with other molecules. All these questions require further investigations in preclinical and *in vitro* models.

Regarding clinical studies, it is important to specify that the anti-inflammatory effects of lycopene and/or tomato products have been evaluated only on plasma. Thus, it will be of particular interest to confirm some of these data in tissue biopsies (notably from adipose tissue) to evaluate the contribution of lycopene or tomato products on tissue inflammation at a local level.

Finally, if the NF-κB signaling pathway is highly suspected to be deactivated by lycopene or its metabolites, the precise molecular mechanism leading to this anti-inflammatory effect is presently not elucidated. Several mechanisms could be involved, such as an effect of the expression of phosphatases involved in the dephosphorylation of NF-κB proteins, physical interaction with NF-kB signaling proteins, or effect on the production of lipid mediators such as resolvins. The fact that lycopene or its metabolites transactivated RAR (Aydemir et al. 2012; Gouranton, Aydemir et al. 2011) also suggests that this nuclear receptor could be involved in lycopene-dependent anti-inflammatory effect since we recently reported that all-*trans* retinoic acid blunted the inflammation process in adipocytes (Karkeni et al. 2016) via an induction of PGC1α, which reduces NF-κB activation and transcriptional control of inflammation (Eisele et al. 2013). The transactivation of other nuclear receptors such as PPARγ by lycopene or its metabolites (Ip et al. 2015) could also be responsible for the anti-inflammatory effect, as previously reported (Szeles, Torocsik, and Nagy 2007).

In conclusion, the impact of lycopene and tomato products on the control of inflammation associated with cardiometabolic disorders is presently well established in preclinical and *in vitro* studies. Further investigations are mandatory to demonstrate the interest of such a nutritional approach in clinical studies. If confirmed, this kind of supplementation could constitute a new strategy to limit or prevent inflammation and its associated consequences.

LIST OF ABBREVIATIONS

AKT	protein kinase B
BCO1	beta-carotene oxygenase 1
BCO2	beta-carotene oxygenase 2
CCL2	chemokine (C-C motif) ligand 2
CCL5	chemokine (C-C motif) ligand 5
CRP	C-reactive protein
CVD	cardiovascular disease
CXCL	chemokine (C-X-C motif) ligand
ERK	extracellular signal-regulated kinase
HF/HChol	high fat/high cholesterol
HMGB1	high-mobility group box 1
HUVEC	human umbilical vein endothelial cell
ICAM	intercellular adhesion molecules
I-κB	inhibitor of nuclear factor kappa B
IL	interleukin
JNK	c-Jun N-terminal kinase
LDL	low-density lipoprotein
LPS	lipopolysaccharide
MCP-1	monocyte chemoattractant protein 1
MMP9	matrix metallopeptidase 9 (MMP9)
NF-κB	nuclear factor kappa B
NHANES III	third national health and nutrition examination survey
NRF2	nuclear factor (erythroid-derived 2)-like 2
PBMC	peripheral blood mononuclear cell
PGC-1α	peroxisome proliferator-activated receptor gamma coactivator 1-α
PI3K	phosphatidylinositol-3-kinase
PPAR	peroxisome proliferator-activated receptor
RANTES	regulated on activation, normal T-cell expressed and secreted
RAR	retinoic acid receptor
SAA	serum amyloid A
SIRT1	Sirtuin 1
THP-1	Tohoku hospital pediatrics-1
TNFα	tumor necrosis factor α

TP   tomato powder
VCAM   vascular cell adhesion molecule
YALTA   young adults longitudinal trends in antioxidants

REFERENCES

Amengual, J., G. P. Lobo, M. Golczak, H. N. Li, T. Klimova, C. L. Hoppel, A. Wyss, K. Palczewski, and J. von Lintig. 2011. "A mitochondrial enzyme degrades carotenoids and protects against oxidative stress." *Faseb J* 25 (3):948–59.

Amengual, J., M. A. Widjaja-Adhi, S. Rodriguez-Santiago, S. Hessel, M. Golczak, K. Palczewski, and J. von Lintig. 2013. "Two carotenoid oxygenases contribute to mammalian provitamin A metabolism." *J Biol Chem* 288 (47):34081–96.

Armoza, A., Y. Haim, A. Bashiri, T. Wolak, and E. Paran. 2013. "Tomato extract and the carotenoids lycopene and lutein improve endothelial function and attenuate inflammatory NF-kappaB signaling in endothelial cells." *J Hypertens* 31 (3):521–9; discussion 529.

Aydemir, G., H. Carlsen, R. Blomhoff, and R. Ruhl. 2012. "Lycopene induces retinoic acid receptor transcriptional activation in mice." *Mol Nutr Food Res* 56 (5):702–12.

Bahcecioglu, I. H., N. Kuzu, K. Metin, I. H. Ozercan, B. Ustundag, K. Sahin, and O. Kucuk. 2010. "Lycopene prevents development of steatohepatitis in experimental nonalcoholic steatohepatitis model induced by high-fat diet." *Vet Med Int* 2010 Oct 3;2010. pii: 262179.

Bakker, G. C., M. J. van Erk, L. Pellis, S. Wopereis, C. M. Rubingh, N. H. Cnubben, T. Kooistra, B. van Ommen, and H. F. Hendriks. 2010. "An antiinflammatory dietary mix modulates inflammation and oxidative and metabolic stress in overweight men: A nutrigenomics approach." *Am J Clin Nutr* 91 (4):1044–59.

Bansal, P., S. K. Gupta, S. K. Ojha, M. Nandave, R. Mittal, S. Kumari, and D. S. Arya. 2006. "Cardioprotective effect of lycopene in the experimental model of myocardial ischemia-reperfusion injury." *Mol Cell Biochem* 289 (1–2):1–9.

Bernal, C., G. Martin-Pozuelo, A. B. Lozano, A. Sevilla, J. Garcia-Alonso, M. Canovas, and M. J. Periago. 2013. "Lipid biomarkers and metabolic effects of lycopene from tomato juice on liver of rats with induced hepatic steatosis." *J Nutr Biochem* 24 (11):1870–81.

Biddle, M. J., T. A. Lennie, G. V. Bricker, R. E. Kopec, S. J. Schwartz, and D. K. Moser. 2014. "Lycopene dietary intervention: A pilot study in patients with heart failure." *J Cardiovasc Nurs* 30 (3):205–12.

Blum, A., M. Monir, K. Khazim, A. Peleg, and N. Blum. 2007. "Tomato-rich (Mediterranean) diet does not modify inflammatory markers." *Clin Invest Med* 30 (2):E70–4.

Bonet, M. L., J. A. Canas, J. Ribot, and A. Palou. 2015. "Carotenoids and their conversion products in the control of adipocyte function, adiposity and obesity." *Arch Biochem Biophys* 572:112–25.

Briviba, K., S. E. Kulling, J. Moseneder, B. Watzl, G. Rechkemmer, and A. Bub. 2004. "Effects of supplementing a low-carotenoid diet with a tomato extract for 2 weeks on endogenous levels of DNA single strand breaks and immune functions in healthy non-smokers and smokers." *Carcinogenesis* 25 (12):2373–8.

Burton-Freeman, B. and H. D. Sesso. 2014. "Whole food versus supplement: Comparing the clinical evidence of tomato intake and lycopene supplementation on cardiovascular risk factors." *Adv Nutr* 5 (5):457–85.

Burton-Freeman, B., J. Talbot, E. Park, S. Krishnankutty, and I. Edirisinghe. 2012. "Protective activity of processed tomato products on postprandial oxidation and inflammation: A clinical trial in healthy weight men and women." *Mol Nutr Food Res* 56 (4):622–31.

Canene-Adams, K., J. K. Campbell, S. Zaripheh, E. H. Jeffery, and J. W. Erdman, Jr. 2005. "The tomato as a functional food." *J Nutr* 135 (5):1226–30.

Catalano, A., R. E. Simone, A. Cittadini, E. Reynaud, C. Caris-Veyrat, and P. Palozza. 2013. "Comparative antioxidant effects of lycopene, apo-10′-lycopenoic acid and apo-14′-lycopenoic acid in human macrophages exposed to H2O2 and cigarette smoke extract." *Food Chem Toxicol* 51:71–9.

Chen, M. L., Y. H. Lin, C. M. Yang, and M. L. Hu. 2012. "Lycopene inhibits angiogenesis both in vitro and in vivo by inhibiting MMP-2/uPA system through VEGFR2-mediated PI3K-Akt and ERK/p38 signaling pathways." *Mol Nutr Food Res* 56 (6):889–99.

Cheng, H. M., G. Koutsidis, J. K. Lodge, A. Ashor, M. Siervo, and J. Lara. 2017. "Tomato and lycopene supplementation and cardiovascular risk factors: A systematic review and meta-analysis." *Atherosclerosis* 257:100–8.

Chung, H. Y., A. L. Ferreira, S. Epstein, S. A. Paiva, C. Castaneda-Sceppa, and E. J. Johnson. 2009. "Site-specific concentrations of carotenoids in adipose tissue: Relations with dietary and serum carotenoid concentrations in healthy adults." *Am J Clin Nutr* 90 (3):533–9.

Chung, J., K. Koo, F. Lian, K. Q. Hu, H. Ernst, and X. D. Wang. 2012. "Apo-10′-lycopenoic acid, a lycopene metabolite, increases sirtuin 1 mRNA and protein levels and decreases hepatic fat accumulation in ob/ob mice." *J Nutr* 142 (3):405–10.

Dallinga-Thie, G. M., R. Franssen, H. L. Mooij, M. E. Visser, H. C. Hassing, F. Peelman, J. J. Kastelein, M. Peterfy, and M. Nieuwdorp. 2010. "The metabolism of triglyceride-rich lipoproteins revisited: New players, new insight." *Atherosclerosis* 211 (1):1–8. doi: 10.1016/j.atherosclerosis.2009.12.027.

dela Sena, C., S. Narayanasamy, K. M. Riedl, R. W. Curley, Jr., S. J. Schwartz, and E. H. Harrison. 2013. "Substrate specificity of purified recombinant human beta-carotene 15,15′-oxygenase (BCO1)." *J Biol Chem* 288 (52):37094–103.

Denniss, S. G., T. D. Haffner, J. T. Kroetsch, S. R. Davidson, J. W. Rush, and R. L. Hughson. 2008. "Effect of short-term lycopene supplementation and postprandial dyslipidemia on plasma antioxidants and biomarkers of endothelial health in young, healthy individuals." *Vasc Health Risk Manag* 4 (1):213–22.

Deplanque, X., D. Muscente-Paque, and E. Chappuis. 2016. "Proprietary tomato extract improves metabolic response to high-fat meal in healthy normal weight subjects." *Food Nutr Res* 60:32537.

Di Tomo, P., R. Canali, D. Ciavardelli, S. Di Silvestre, A. De Marco, A. Giardinelli, C. Pipino, N. Di Pietro, F. Virgili, and A. Pandolfi. 2012. "beta-Carotene and lycopene affect endothelial response to TNF-alpha reducing nitro-oxidative stress and interaction with monocytes." *Mol Nutr Food Res* 56 (2):217–27.

Eisele, P. S., S. Salatino, J. Sobek, M. O. Hottiger, and C. Handschin. 2013. "The peroxisome proliferator-activated receptor gamma coactivator 1alpha/beta (PGC-1) coactivators repress the transcriptional activity of NF-kappaB in skeletal muscle cells." *J Biol Chem* 288 (4):2246–60.

Engelmann, N. J., S. K. Clinton, and J. W. Erdman, Jr. 2011. "Nutritional aspects of phytoene and phytofluene, carotenoid precursors to lycopene." *Adv Nutr* 2 (1):51–61. doi: 10.3945/an.110.000075.

Esser, N., N. Paquot, and A. J. Scheen. 2015. "Inflammatory markers and cardiometabolic diseases." *Acta Clin Belg* 70 (3):193–9. doi: 10.1179/2295333715Y.0000000004.

Feng, D., W. H. Ling, and R. D. Duan. 2010. "Lycopene suppresses LPS-induced NO and IL-6 production by inhibiting the activation of ERK, p38MAPK, and NF-kappaB in macrophages." *Inflamm Res* 59 (2):115–21.

Fenni, S., H. Hammou, J. Astier, L. Bonnet, E. Karkeni, C. Couturier, F. Tourniaire, and J. F. Landrier. 2017. "Lycopene and tomato powder supplementation similarly inhibit high-fat diet induced obesity, inflammatory response, and associated metabolic disorders." *Mol Nutr Food Res* 2017 Sep;61(9). doi: 10.1002/mnfr.201601083. Epub 2017 Apr 7.

Ford, E. S., S. Liu, D. M. Mannino, W. H. Giles, and S. J. Smith. 2003. "C-reactive protein concentration and concentrations of blood vitamins, carotenoids, and selenium among United States adults." *Eur J Clin Nutr* 57 (9):1157–63.

Ford, N. A., S. K. Clinton, J. von Lintig, A. Wyss, and J. W. Erdman, Jr. 2010. "Loss of carotene-9′,10′-monooxygenase expression increases serum and tissue lycopene concentrations in lycopene-fed mice." *J Nutr* 140 (12):2134–8.

Ford, N. A., A. C. Elsen, and J. W. Erdman, Jr. 2013. "Genetic ablation of carotene oxygenases and consumption of lycopene or tomato powder diets modulate carotenoid and lipid metabolism in mice." *Nutr Res* 33 (9):733–42.

Gajendragadkar, P. R., A. Hubsch, K. M. Maki-Petaja, M. Serg, I. B. Wilkinson, and J. Cheriyan. 2014. "Effects of oral lycopene supplementation on vascular function in patients with cardiovascular disease and healthy volunteers: A randomised controlled trial." *PLoS One* 9 (6):e99070.

Gajic, M., S. Zaripheh, F. Sun, and J. W. Erdman, Jr. 2006. "Apo-8′-lycopenal and apo-12′-lycopenal are metabolic products of lycopene in rat liver." *J Nutr* 136 (6):1552–7.

Ghavipour, M., A. Saedisomeolia, M. Djalali, G. Sotoudeh, M. R. Eshraghyan, A. M. Moghadam, and L. G. Wood. 2012. "Tomato juice consumption reduces systemic inflammation in overweight and obese females." *Br J Nutr* 109 (11):2031–5.

Gouranton, E., G. Aydemir, E. Reynaud, J. Marcotorchino, C. Malezet, C. Caris-Veyrat, R. Blomhoff, J. F. Landrier, and R. Ruhl. 2011. "Apo-10′-lycopenoic acid impacts adipose tissue biology via the retinoic acid receptors." *Biochim Biophys Acta* 1811 (12):1105–14.

Gouranton, E., C. Thabuis, C. Riollet, C. Malezet-Desmoulins, C. El Yazidi, M. J. Amiot, P. Borel, and J. F. Landrier. 2011. "Lycopene inhibits proinflammatory cytokine and chemokine expression in adipose tissue." *J Nutr Biochem* 22 (7):642–8.

Gouranton, E., C. E. Yazidi, N. Cardinault, M. J. Amiot, P. Borel, and J. F. Landrier. 2008. "Purified low-density lipoprotein and bovine serum albumin efficiency to internalise lycopene into adipocytes." *Food Chem Toxicol* 46 (12):3832–6.

Gregor, M. F. and G. S. Hotamisligil. 2011. "Inflammatory mechanisms in obesity." *Annu Rev Immunol* 29:415–45.

Hadad, N. and R. Levy. 2012. "The synergistic anti-inflammatory effects of lycopene, lutein, beta-carotene, and carnosic acid combinations via redox-based inhibition of NF-kappaB signaling." *Free Radic Biol Med* 53 (7):1381–91.

Hazewindus, M., G. R. Haenen, A. R. Weseler, and A. Bast. 2014. "Protection against chemotaxis in the anti-inflammatory effect of bioactives from tomato ketchup." *PLoS One* 9 (12):e114387.

He, Q., W. Zhou, C. Xiong, G. Tan, and M. Chen. 2015. "Lycopene attenuates inflammation and apoptosis in post-myocardial infarction remodeling by inhibiting the nuclear factor-kappaB signaling pathway." *Mol Med Rep* 11 (1):374–8.

Hozawa, A., D. R. Jacobs, Jr., M. W. Steffes, M. D. Gross, L. M. Steffen, and D. H. Lee. 2007. "Relationships of circulating carotenoid concentrations with several markers of inflammation, oxidative stress, and endothelial dysfunction: The Coronary Artery Risk Development in Young Adults (CARDIA)/Young Adult Longitudinal Trends in Antioxidants (YALTA) study." *Clin Chem* 53 (3):447–55.

Hu, K. Q., C. Liu, H. Ernst, N. I. Krinsky, R. M. Russell, and X. D. Wang. 2006. "The biochemical characterization of ferret carotene-9′,10′-monooxygenase catalyzing cleavage of carotenoids in vitro and in vivo." *J Biol Chem* 281 (28):19327–38.

Hung, C. F., T. F. Huang, B. H. Chen, J. M. Shieh, P. H. Wu, and W. B. Wu. 2008. "Lycopene inhibits TNF-alpha-induced endothelial ICAM-1 expression and monocyte-endothelial adhesion." *Eur J Pharmacol* 586 (1–3):275–82.

Ip, B. C., K. Q. Hu, C. Liu, D. E. Smith, M. S. Obin, L. M. Ausman, and X. D. Wang. 2014. "Lycopene metabolite, apo-10′-lycopenoic acid, inhibits diethylnitrosamine-initiated, high fat diet-promoted hepatic inflammation and tumorigenesis in mice." *Cancer Prev Res (Phila)* 6 (12):1304–16.

Ip, B. C., C. Liu, A. H. Lichtenstein, J. von Lintig, and X. D. Wang. 2015. "Lycopene and apo-10′-lycopenoic acid have differential mechanisms of protection against hepatic steatosis in beta-carotene-9′,10′-oxygenase knockout male mice." *J Nutr* 145 (2):268–76.

Jacob, K., M. J. Periago, V. Bohm, and G. R. Berruezo. 2008. "Influence of lycopene and vitamin C from tomato juice on biomarkers of oxidative stress and inflammation." *Br J Nutr* 99 (1):137–46.

Karkeni, E., J. Astier, F. Tourniaire, M. El Abed, B. Romier, E. Gouranton, L. Wan, P. Borel, J. Salles, S. Walrand, J. Ye, and J. F. Landrier. 2016. "Obesity-associated inflammation induces microRNA-155 expression in adipocytes and adipose tissue: Outcome on adipocyte function." *J Clin Endocrinol Metab* 101 (4):1615–26. doi: 10.1210 /jc.2015-3410.

Khachik, F., F. F. de Moura, D. Y. Zhao, C. P. Aebischer, and P. S. Bernstein. 2002. "Transformations of selected carotenoids in plasma, liver, and ocular tissues of humans and in nonprimate animal models." *Invest Ophthalmol Vis Sci* 43 (11):3383–92.

Khachik, F., A. Steck, U. A. Niggli, and H. Pfander. 1998. "Partial synthesis and structural elucidation of the oxidative metabolites of lycopene identified in tomato paste, tomato juice, and human serum." *J Agric Food Chem* 46 (12):4874–84.

Kiefer, C., S. Hessel, J. M. Lampert, K. Vogt, M. O. Lederer, D. E. Breithaupt, and J. von Lintig. 2001. "Identification and characterization of a mammalian enzyme catalyzing the asymmetric oxidative cleavage of provitamin A." *J Biol Chem* 276 (17):14110–16.

Kim, J. Y., J. K. Paik, O. Y. Kim, H. W. Park, J. H. Lee, Y. Jang, and J. H. Lee. 2011. "Effects of lycopene supplementation on oxidative stress and markers of endothelial function in healthy men." *Atherosclerosis* 215 (1):189–95.

Kim, O. Y., H. Y. Yoe, H. J. Kim, J. Y. Park, J. Y. Kim, S. H. Lee, J. H. Lee, K. P. Lee, Y. Jang, and J. H. Lee. 2010. "Independent inverse relationship between serum lycopene concentration and arterial stiffness." *Atherosclerosis* 208 (2):581–6.

Kim, Y. I., S. Mohri, S. Hirai, T. Lin, T. Goto, C. Ohyane, T. Sakamoto, H. Takahashi, D. Shibata, N. Takahashi, and T. Kawada. 2015. "Tomato extract suppresses the production of proinflammatory mediators induced by interaction between adipocytes and macrophages." *Biosci Biotechnol Biochem* 79 (1):82–7.

Kopec, R. E., K. M. Riedl, E. H. Harrison, R. W. Curley, Jr., D. P. Hruszkewycz, S. K. Clinton, and S. J. Schwartz. 2010. "Identification and quantification of apo-lycopenals in fruits, vegetables, and human plasma." *J Agric Food Chem* 58 (6):3290–6.

Kritchevsky, S. B., A. J. Bush, M. Pahor, and M. D. Gross. 2000. "Serum carotenoids and markers of inflammation in nonsmokers." *Am J Epidemiol* 152 (11):1065–71.

Landrier, J. F., J. Marcotorchino, and F. Tourniaire. 2012. "Lipophilic micronutrients and adipose tissue biology." *Nutrients* 4 (11):1622–49.

Lee, W., S. K. Ku, J. W. Bae, and J. S. Bae. 2012. "Inhibitory effects of lycopene on HMGB1-mediated pro-inflammatory responses in both cellular and animal models." *Food Chem Toxicol* 50 (6):1826–33.

Li, Y. F., Y. Y. Chang, H. C. Huang, Y. C. Wu, M. D. Yang, and P. M. Chao. 2015. "Tomato juice supplementation in young women reduces inflammatory adipokine levels independently of body fat reduction." *Nutrition* 31 (5):691–6.

Liu, X., D. Qu, F. He, Q. Lu, J. Wang, and D. Cai. 2007. "Effect of lycopene on the vascular endothelial function and expression of inflammatory agents in hyperhomocysteinemic rats." *Asia Pac J Clin Nutr* 16 Suppl 1:244–8.

Luvizotto Rde, A., A. F. Nascimento, E. Imaizumi, D. T. Pierine, S. J. Conde, C. R. Correa, K. J. Yeum, and A. L. Ferreira. 2013. "Lycopene supplementation modulates plasma concentrations and epididymal adipose tissue mRNA of leptin, resistin and IL-6 in diet-induced obese rats." *Br J Nutr* 110 (10):1803–9.

Marcotorchino, J., B. Romier, E. Gouranton, C. Riollet, B. Gleize, C. Malezet-Desmoulins, and J. F. Landrier. 2012. "Lycopene attenuates LPS-induced TNF-alpha secretion in macrophages and inflammatory markers in adipocytes exposed to macrophage-conditioned media." *Mol Nutr Food Res* 56 (5):725–32.

Markovits, N., A. Ben Amotz, and Y. Levy. 2009. "The effect of tomato-derived lycopene on low carotenoids and enhanced systemic inflammation and oxidation in severe obesity." *Isr Med Assoc J* 11 (10):598–601.

Martin-Pozuelo, G., I. Navarro-Gonzalez, R. Gonzalez-Barrio, M. Santaella, J. Garcia-Alonso, N. Hidalgo, C. Gomez-Gallego, G. Ros, and M. J. Periago. 2015. "The effect of tomato juice supplementation on biomarkers and gene expression related to lipid metabolism in rats with induced hepatic steatosis." *Eur J Nutr* 54 (6):933–44.

McEneny, J., L. Wade, I. S. Young, L. Masson, G. Duthie, A. McGinty, C. McMaster, and F. Thies. 2013. "Lycopene intervention reduces inflammation and improves HDL functionality in moderately overweight middle-aged individuals." *J Nutr Biochem* 24 (1):163–8.

Moran, N. E., M. J. Cichon, K. M. Riedl, E. M. Grainger, S. J. Schwartz, J. A. Novotny, J. W. Erdman, Jr., and S. K. Clinton. 2015. "Compartmental and noncompartmental modeling of (1)(3)C-lycopene absorption, isomerization, and distribution kinetics in healthy adults." *Am J Clin Nutr* 102 (6):1436–49.

Moran, N. E., J. W. Erdman, Jr., and S. K. Clinton. 2013. "Complex interactions between dietary and genetic factors impact lycopene metabolism and distribution." *Arch Biochem Biophys* 539 (2):171–80.

Moussa, M., E. Gouranton, B. Gleize, C. E. Yazidi, I. Niot, P. Besnard, P. Borel, and J. F. Landrier. 2011. "CD36 is involved in lycopene and lutein uptake by adipocytes and adipose tissue cultures." *Mol Nutr Food Res* 55 (4):578–84.

Muller, L., C. Caris-Veyrat, G. Lowe, and V. Bohm. 2015. "Lycopene and its antioxidant role in the prevention of cardiovascular diseases—A critical review." *Crit Rev Food Sci Nutr* 56 (11):1868–79.

Navarrete, S., M. Alarcon, and I. Palomo. 2015. "Aqueous extract of tomato (Solanum lycopersicum L.) and ferulic acid reduce the expression of TNF-alpha and IL-1beta in LPS-activated macrophages." *Molecules* 20 (8):15319–29.

Palczewski, G., J. Amengual, C. L. Hoppel, and J. von Lintig. 2014. "Evidence for compartmentalization of mammalian carotenoid metabolism." *Faseb J* 28 (10):4457–69.

Palozza, P., N. Parrone, A. Catalano, and R. Simone. 2010. "Tomato lycopene and inflammatory cascade: Basic interactions and clinical implications." *Curr Med Chem* 17 (23):2547–63.

Palozza, P., R. E. Simone, A. Catalano, F. Saraceni, L. Celleno, M. C. Mele, G. Monego, and A. Cittadini. 2012. "Modulation of MMP-9 pathway by lycopene in macrophages and fibroblasts exposed to cigarette smoke." *Inflamm Allergy Drug Targets* 11 (1):36–47.

Pearson, T. A., G. A. Mensah, R. W. Alexander, J. L. Anderson, R. O. Cannon, 3rd, M. Criqui, Y. Y. Fadl, S. P. Fortmann, Y. Hong, G. L. Myers, N. Rifai, S. C. Smith, Jr., K. Taubert, R. P. Tracy, and F. Vinicor. 2003. "Markers of inflammation and cardiovascular disease: Application to clinical and public health practice: A statement for healthcare professionals from the Centers for Disease Control and Prevention and the American Heart Association." *Circulation* 107 (3):499–511.

Petyaev, I. M., P. Y. Dovgalevsky, V. A. Klochkov, N. E. Chalyk, and N. Kyle. 2012. "Whey protein lycosome formulation improves vascular functions and plasma lipids with reduction of markers of inflammation and oxidative stress in prehypertension." *ScientificWorldJournal* 2012:269476.

Pierine, D. T., M. E. Navarro, I. O. Minatel, R. A. Luvizotto, A. F. Nascimento, A. L. Ferreira, K. J. Yeum, and C. R. Correa. 2014. "Lycopene supplementation reduces TNF-alpha via RAGE in the kidney of obese rats." *Nutr Diabetes* 4:e142.

Raghuvanshi, S., V. Reed, W. S. Blaner, and E. H. Harrison. 2015. "Cellular localization of beta-carotene 15,15′ oxygenase-1 (BCO1) and beta-carotene 9′,10′ oxygenase-2 (BCO2) in rat liver and intestine." *Arch Biochem Biophys* 572:19–27.

Rao, A. V., M. R. Ray, and L. G. Rao. 2006. "Lycopene." *Adv Food Nutr Res* 51:99–164.

Redmond, T. M., S. Gentleman, T. Duncan, S. Yu, B. Wiggert, E. Gantt, and F. X. Cunningham, Jr. 2001. "Identification, expression, and substrate specificity of a mammalian beta-carotene 15,15′-dioxygenase." *J Biol Chem* 276 (9):6560–5.

Renju, G. L., G. M. Kurup, and C. H. Saritha Kumari. 2014. "Effect of lycopene from Chlorella marina on high cholesterol-induced oxidative damage and inflammation in rats." *Inflammopharmacology* 22 (1):45–54.

Richelle, M., P. Lambelet, A. Rytz, I. Tavazzi, A. F. Mermoud, C. Juhel, P. Borel, and K. Bortlik. 2012. "The proportion of lycopene isomers in human plasma is modulated by lycopene isomer profile in the meal but not by lycopene preparation." *Br J Nutr* 107 (10):1482–8. doi: 10.1017/S0007114511004569.

Richelle, M., B. Sanchez, I. Tavazzi, P. Lambelet, K. Bortlik, and G. Williamson. 2010. "Lycopene isomerisation takes place within enterocytes during absorption in human subjects." *Br J Nutr* 103 (12):1800–7. doi: 10.1017/S0007114510000103.

Ridker, P. M. and D. A. Morrow. 2003. "C-reactive protein, inflammation, and coronary risk." *Cardiol Clin* 21 (3):315–25.

Riso, P., F. Visioli, S. Grande, S. Guarnieri, C. Gardana, P. Simonetti, and M. Porrini. 2006. "Effect of a tomato-based drink on markers of inflammation, immunomodulation, and oxidative stress." *J Agric Food Chem* 54 (7):2563–6.

Ross, A. B., T. Vuong le, J. Ruckle, H. A. Synal, T. Schulze-Konig, K. Wertz, R. Rumbeli, R. G. Liberman, P. L. Skipper, S. R. Tannenbaum, A. Bourgeois, P. A. Guy, M. Enslen, I. L. Nielsen, S. Kochhar, M. Richelle, L. B. Fay, and G. Williamson. 2011. "Lycopene bioavailability and metabolism in humans: An accelerator mass spectrometry study." *Am J Clin Nutr* 93 (6):1263–73.

Rowley, K., K. Z. Walker, J. Cohen, A. J. Jenkins, D. O'Neal, Q. Su, J. D. Best, and K. O'Dea. 2003. "Inflammation and vascular endothelial activation in an Aboriginal population: Relationships to coronary disease risk factors and nutritional markers." *Med J Aust* 178 (10):495–500.

Ruhl, R., and J. F. Landrier. 2015. "Dietary regulation of adiponectin by direct and indirect lipid activators of nuclear hormone receptors." *Mol Nutr Food Res* 60 (1):175–84.

Ryden, M., P. Garvin, M. Kristenson, P. Leanderson, J. Ernerudh, and L. Jonasson. 2012. "Provitamin A carotenoids are independently associated with matrix metalloproteinase-9 in plasma samples from a general population." *J Intern Med* 272 (4):371–84.

Sanchez-Moreno, C., M. P. Cano, B. de Ancos, L. Plaza, B. Olmedilla, F. Granado, and A. Martin. 2006. "Mediterranean vegetable soup consumption increases plasma vitamin C and decreases F2-isoprostanes, prostaglandin E2 and monocyte chemotactic protein-1 in healthy humans." *J Nutr Biochem* 17 (3):183–9.

Schug, T. T., and X. Li. 2011. "Sirtuin 1 in lipid metabolism and obesity." *Ann Med* 43 (3):198–211.

Schwager, J., N. Richard, B. Mussler, and D. Raederstorff. 2016. "Tomato aqueous extract modulates the inflammatory profile of immune cells and endothelial cells." *Molecules* 21 (2):168.

Singh, D. P., P. Khare, J. Zhu, K. K. Kondepudi, J. Singh, R. K. Baboota, R. K. Boparai, R. Khardori, K. Chopra, and M. Bishnoi. 2016. "A novel cobiotic-based preventive approach against high-fat diet-induced adiposity, nonalcoholic fatty liver and gut derangement in mice." *Int J Obes (Lond)* 40 (3):487–96.

Story, E. N., R. E. Kopec, S. J. Schwartz, and G. K. Harris. 2010. "An update on the health effects of tomato lycopene." *Annu Rev Food Sci Technol* 1:189–210.

Sung, L. C., H. H. Chao, C. H. Chen, J. C. Tsai, J. C. Liu, H. J. Hong, T. H. Cheng, and J. J. Chen. 2015. "Lycopene inhibits cyclic strain-induced endothelin-1 expression through the suppression of reactive oxygen species generation and induction of heme oxygenase-1 in human umbilical vein endothelial cells." *Clin Exp Pharmacol Physiol* 42 (6):632–9.

Sy, C., B. Gleize, O. Dangles, J. F. Landrier, C. C. Veyrat, and P. Borel. 2012. "Effects of physicochemical properties of carotenoids on their bioaccessibility, intestinal cell uptake, and blood and tissue concentrations." *Mol Nutr Food Res* 2012 Sep;56 (9):1385–97. doi: 10.1002/mnfr.201200041. Epub 2012 Jul 20.

Szeles, L., D. Torocsik, and L. Nagy. 2007. "PPARgamma in immunity and inflammation: Cell types and diseases." *Biochim Biophys Acta* 1771 (8):1014–30.

Tan, H. L., N. E. Moran, M. J. Cichon, K. M. Riedl, S. J. Schwartz, J. W. Erdman, Jr., D. K. Pearl, J. M. Thomas-Ahner, and S. K. Clinton. 2014. "beta-Carotene-9′,10′-oxygenase status modulates the impact of dietary tomato and lycopene on hepatic nuclear receptor-, stress-, and metabolism-related gene expression in mice." *J Nutr* 144 (4):431–9.

Thies, F., L. F. Masson, A. Rudd, N. Vaughan, C. Tsang, J. Brittenden, W. G. Simpson, S. Duthie, G. W. Horgan, and G. Duthie. 2012. "Effect of a tomato-rich diet on markers of cardiovascular disease risk in moderately overweight, disease-free, middle-aged adults: A randomized controlled trial." *Am J Clin Nutr* 95 (5):1013–22.

Tong, C., C. Peng, L. Wang, L. Zhang, X. Yang, P. Xu, J. Li, T. Delplancke, H. Zhang, and H. Qi. 2016. "Intravenous administration of lycopene, a tomato extract, protects against myocardial ischemia-reperfusion injury." *Nutrients* 8 (3):138.

Tourniaire, F., B. Romier-Crouzet, J. H. Lee, J. Marcotorchino, E. Gouranton, J. Salles, C. Malezet, J. Astier, P. Darmon, E. Blouin, S. Walrand, J. Ye, and J. F. Landrier. 2013. "Chemokine expression in inflamed adipose tissue is mainly mediated by NF-kappaB." *PLoS One* 8 (6):e66515.

Tsitsimpikou, C., K. Tsarouhas, N. Kioukia-Fougia, C. Skondra, P. Fragkiadaki, P. Papalexis, P. Stamatopoulos, I. Kaplanis, A. W. Hayes, A. Tsatsakis, and E. Rentoukas. 2014. "Dietary supplementation with tomato-juice in patients with metabolic syndrome: A suggestion to alleviate detrimental clinical factors." *Food Chem Toxicol* 74:9–13.

Upaganlawar, A., H. Gandhi, and R. Balaraman. 2010. "Effect of vitamin E alone and in combination with lycopene on biochemical and histopathological alterations in isoproterenol-induced myocardial infarction in rats." *J Pharmacol Pharmacother* 1 (1):24–31. doi: 10.4103/0976-500X.64532.

Upritchard, J. E., W. H. Sutherland, and J. I. Mann. 2000. "Effect of supplementation with tomato juice, vitamin E, and vitamin C on LDL oxidation and products of inflammatory activity in type 2 diabetes." *Diabetes Care* 23 (6):733–8.

Valderas-Martinez, P., G. Chiva-Blanch, R. Casas, S. Arranz, M. Martinez-Huelamo, M. Urpi-Sarda, X. Torrado, D. Corella, R. M. Lamuela-Raventos, and R. Estruch. 2016. "Tomato sauce enriched with olive oil exerts greater effects on cardiovascular disease risk factors than raw tomato and tomato sauce: A randomized trial." *Nutrients* 8 (3):170.

van Herpen-Broekmans, W. M., I. A. Klopping-Ketelaars, M. L. Bots, C. Kluft, H. Princen, H. F. Hendriks, L. B. Tijburg, G. van Poppel, and A. F. Kardinaal. 2004. "Serum carotenoids and vitamins in relation to markers of endothelial function and inflammation." *Eur J Epidemiol* 19 (10):915–21.

Walston, J., Q. Xue, R. D. Semba, L. Ferrucci, A. R. Cappola, M. Ricks, J. Guralnik, and L. P. Fried. 2006. "Serum antioxidants, inflammation, and total mortality in older women." *Am J Epidemiol* 163 (1):18–26.

Wang, L., J. M. Gaziano, E. P. Norkus, J. E. Buring, and H. D. Sesso. 2008. "Associations of plasma carotenoids with risk factors and biomarkers related to cardiovascular disease in middle-aged and older women." *Am J Clin Nutr* 88 (3):747–54.

Wang, X. D. 2012. "Lycopene metabolism and its biological significance." *Am J Clin Nutr* 96 (5):1214S–22S.

Wang, X., H. Lv, Y. Gu, X. Wang, H. Cao, Y. Tang, H. Chen, and C. Huang. 2014. "Protective effect of lycopene on cardiac function and myocardial fibrosis after acute myocardial infarction in rats via the modulation of p38 and MMP-9." *J Mol Histol* 45 (1):113–20.

Wang, Y., Y. Gao, W. Yu, Z. Jiang, J. Qu, and K. Li. 2013. "Lycopene protects against LPS-induced proinflammatory cytokine cascade in HUVECs." *Pharmazie* 68 (8):681–4.

Warnke, I., J. W. Jocken, R. Schoop, C. Toepfer, R. Goralczyk, and J. Schwager. 2016. "Combinations of bio-active dietary constituents affect human white adipocyte function in-vitro." *Nutr Metab (Lond)* 13:84.

Watzl, B., A. Bub, K. Briviba, and G. Rechkemmer. 2003. "Supplementation of a low-carotenoid diet with tomato or carrot juice modulates immune functions in healthy men." *Ann Nutr Metab* 47 (6):255–61.

Xu, J., H. Hu, B. Chen, R. Yue, Z. Zhou, Y. Liu, S. Zhang, L. Xu, H. Wang, and Z. Yu. 2015. "Lycopene protects against hypoxia/reoxygenation injury by alleviating ER stress induced apoptosis in neonatal mouse cardiomyocytes." *PLoS One* 10 (8):e0136443.

Yeon, J. Y., H. S. Kim, and M. K. Sung. 2012. "Diets rich in fruits and vegetables suppress blood biomarkers of metabolic stress in overweight women." *Prev Med* 54 Suppl:S109–15.

Yin, Q., Y. Ma, Y. Hong, X. Hou, J. Chen, C. Shen, M. Sun, Y. Shang, S. Dong, Z. Zeng, J. J. Pei, and X. Liu. 2014. "Lycopene attenuates insulin signaling deficits, oxidative stress, neuroinflammation, and cognitive impairment in fructose-drinking insulin resistant rats." *Neuropharmacology* 86:389–96.

Zidani, S., A. Benakmoum, A. Ammouche, Y. Benali, A. Bouhadef, and S. Abbeddou. 2017. "Effect of dry tomato peel supplementation on glucose tolerance, insulin resistance, and hepatic markers in mice fed high-saturated-fat/high-cholesterol diets." *J Nutr Biochem* 40:164–71.

Zou, J., D. Feng, W. H. Ling, and R. D. Duan. 2013. "Lycopene suppresses proinflammatory response in lipopolysaccharide-stimulated macrophages by inhibiting ROS-induced trafficking of TLR4 to lipid raft-like domains." *J Nutr Biochem* 24 (6):1117–22.

6 Lycopene, Tomatoes, and Bone Health

Leticia G. Rao and A. Venketeshwer Rao

CONTENTS

6.1 INTRODUCTION

6.1.1 OSTEOPOROSIS

Osteoporosis is a metabolic bone disease characterized by gradual bone loss over many years. It is known as the "silent thief" since there are generally no noticeable symptoms until the fragile bones result in fracture [1]. The two major bone cells, the bone-forming osteoblasts and the bone-resorbing osteoclasts, are involved in the process of bone remodeling that characterizes bone as a dynamic tissue that continuously renews throughout life. The coupled process of bone formation and bone resorption during remodelling in healthy bone is tightly regulated. Metabolic bone diseases result from the disturbances in the remodeling process [2]. Some of the risk factors for osteoporosis known to disturb the remodeling process [3,4] are presented in Table 6.1. The risk factors that are of interest in this review are oxidative stress-generating factors, including smoking, alcohol intake, low antioxidant status, nutrition deficiency, excessive sports activity, and excessive caffeine intake, all of which having been shown to increase the rate of bone loss that eventually leads to osteoporosis.

TABLE 6.1

Risk Factors for Osteoporosis

Unmodifiable	Modifiable
Race	Chronic inactivity
Sex	Low body weight
Age	Low lifetime calcium intake
Genetics	Medication used
Body size	Oxidative stress-related factors
Family history	Smoking
Previous fractures	Alcohol intake
	Low antioxidant status
	Nutrition deficiency
	Excessive sports activity
	Excessive caffeine intake

Source: Reproduced from Rao, L. G., and A. V. Rao. "Oxidative Stress and Antioxidants in the Risk of Osteoporosis—Role of the Antioxidants Lycopene and Polyphenols," in: *Topics in Osteoporosis*, ed. Margarita Valdes Flores, 2013. Available at http://www.intechopen.com/articles/show/title/oxidative-stress-and-antioxidants-in-the-risk-of-osteoporosis-role-of-the-anti oxidants-lycopene-and-. ISBN: 978-953-51-1066-8. With permission from Intech.

6.1.2 TREATMENT OF OSTEOPOROSIS

Advances in the knowledge on osteoporosis is not without pitfalls. Once a first line of treatment for osteoporosis, Hormone Replacement Therapy (HRT), has been discontinued due to side effects [4]. Several drugs, known as bisphosphonates, in spite of being effective in stopping the resorption of bone and preventing osteoporosis in postmenopausal women, are now recognized as having a number of side effects [5–7]. The alarming side effects have been causing a number of women with osteoporosis to resort to other modes of treatment, including to those of natural food components.

6.1.3 COMPLEMENTARY AND ALTERNATIVE APPROACH TO PREVENTION AND TREATMENT OF OSTEOPOROSIS

There is an increasing demand for complementary and alternative medicine (CAM) for the prevention and treatment of osteoporosis [8] due to the adverse side effects of HRT and the bisphosphonates in the management of postmenopausal osteoporosis. CAM is the term for medical services, practices, and products that are different from standard care. These include diet, acupuncture, exercise, and nutritional supplements. Recent dietary guidelines worldwide for the prevention of chronic diseases include an increase in the consumption of fruit and vegetables [9] that are good sources of dietary antioxidants [10]. Epidemiologically and clinical intervention have demonstrated the beneficial effects of antioxidants in bone health and osteoporosis. There is now strong scientific support for the potential benefits of incorporating therapeutic nutritional interventions with contemporary pharmaceutical treatments [11] since many nutrients have been identified as being beneficial to bone health [12,13]. We will review in this chapter our clinical studies on lycopene treatment that showed positive results on bone health.

6.1.4 OXIDATIVE STRESS AND ANTIOXIDANTS

Under normal physiological conditions, oxidative stress can be prevented by promoting antioxidant defenses. Several mechanisms for endogenous defense are present in the body [14]. Exogenous antioxidants from dietary sources present in fruit and vegetables are also utilized to combat oxidative stress [15]. Lycopene, a potent antioxidant [15], is a lipid-soluble carotenoid. Figure 6.1 is a cartoon illustrating the protection afforded by antioxidants when oxidative stress from ROS, environmental pollution, and lifestyle factors can cause the damaging effects on DNA, lipid, and protein that subsequently results in chronic diseases.

6.2 STUDIES ON LYCOPENE

The direct role of lycopene in osteoblasts and osteoclasts reported in the literature no doubt has contributed to opening up research into human studies on the beneficial role of lycopene in the prevention of the risk for osteoporosis. Both epidemiological and clinical intervention studies have further supported this beneficial role of lycopene.

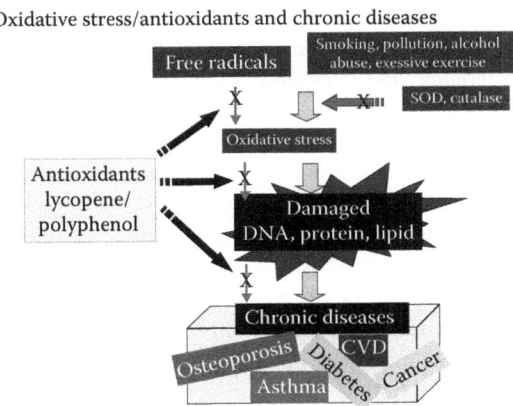

FIGURE 6.1 Cartoon demonstrating the damaging effects of oxidative stress and the beneficial effects of antioxidants to prevent the development of chronic diseases. (Reproduced from Rao, L. G., and A. V. Rao. "Oxidative Stress and Antioxidants in the Risk of Osteoporosis—Role of the Antioxidants Lycopene and Polyphenols," in: *Topics in Osteoporosis*, ed. Margarita Valdes Flores, 2013. Available at http://www.intechopen.com/articles/show/title/oxidative-stress-and-antioxidants-in-the-risk-of-osteoporosis-role-of-the-antioxidants-lycopene-and-. ISBN: 978-953-51-1066-8. With permission from Intech.)

6.2.1 IN VITRO STUDIES OF LYCOPENE IN OSTEOBLASTS AND OSTEOCLASTS

It is not surprising that only few studies on the effects of lycopene in osteoblasts and osteoclasts have been reported. The most likely reason is that lycopene is neither soluble in water nor in the culture medium. Lycopene must be solubilized in organic solvent before it can be added to the cell culture. When Lyc-o-mato preparation from Lyco-red that is partially dispersed in micelle form in water was used by Kim et al. [16], lycopene was shown to have a stimulatory effect on cell proliferation as well as a stimulatory effect on alkaline phosphatase activity in human osteoblast-like SaOS-2 cells in culture. This is the first report on the effect of lycopene on human osteoblasts [16]. Other effects of lycopene on cells in mice have also been reported [17]. Rao et al. [18,19] showed that only the *cis* isomers, and none of the trans isomers of lycopene, are able to prevent and repair the damaging effects of H_2O_2-induced oxidative stress on the formation of mineralized bone nodules in human SaOS-2 cells.

Only two studies on the effects of lycopene in osteoclasts have been reported [20,21]. Lycopene was shown to inhibit multinucleated osteoclast cell formation as well as the formation of ROS-secreting osteoclasts in bone marrow cells from rat femur cultured in cultured in 16-multi-well slide coated with Osteologic™ calcium phosphate [20]. Ishimi et al. [21] also reported the effect of lycopene on osteoclast formation and bone resorption by murine osteoclasts when co-cultured with calvarial osteoblasts.

6.2.2 EPIDEMIOLOGICAL STUDIES ON LYCOPENE

The major component of the Mediterranean diet, a model diet for disease prevention, is abundant plant foods, including fruits and vegetables [22]. There is a possibility that one of the active components is lycopene [23,24]. Epidemiological studies on the prevention of osteoporosis in the Mediterranean population support the beneficial role of tomatoes and tomato products present in the Mediterranean diet [25].

Specific epidemiological and intervention studies on the role of lycopene in osteoporosis risk prevention have been reported and recently reviewed [15,19]. In summary, several epidemiological studies on the role of lycopene as an antioxidant in the prevention of oxidative stress-related osteoporosis and correlation with bone mineral density in postmenopausal women have been reported [19,26–29].

An epidemiological study to determine the beneficial role of lycopene in the prevention of risk for osteoporosis was carried out by Rao et al. [19,30]. A cross-sectional study was carried out in which a total of 33 postmenopausal women from 50 to 60 years old were recruited to participate. A seven-day dietary record and blood samples were provided by the participants. Blood samples were analyzed for bone turnover markers, including the bone formation and bone resorption marker NTX; total antioxidant capacity; and oxidative stress parameters, including lipid peroxidation and protein oxidation. The data were analyzed to find out if there were differences in these parameters between those with lower and higher consumption of lycopene. Their results also showed that the estimated dietary lycopene had a direct and significant correlation with serum lycopene. These results suggest that lycopene from the diet is bioavailable. Rao et al. [30] concluded that a lower protein oxidation ($P < 0.05$) and low NTX ($P < 0.005$) were associated with higher serum lycopene and support the antioxidative properties of lycopene involvement in its mechanisms of action [30].

Thus, the beneficial role of lycopene in the prevention of risk for osteoporosis is supported by our epidemiological studies. Clinical studies described below further support this conclusion.

6.2.3 Animal Intervention Studies on Lycopene

Other than our intervention studies, to be discussed in the next section, most of the intervention studies with lycopene were carried out in ovariectomized (OVXed) rats as models of postmenopausal osteoporosis. Thus, Iimura showed that lycopene intake by OVXed rats facilitates the increase of bone mineral density [31] and that lycopene has effects on serum mineral elements and bone strength in OVXed rats [32]. Adwani et al. [33] showed that lycopene treatment can help against the loss of bone mass, microarchitecture, and strength in relation to regulatory mechanisms in a postmenopausal osteoporosis rat model. Their findings demonstrate that lycopene treatment in OVXed rats primarily suppressed bone turnover to restore bone strength and microarchitecture.

6.2.4 Clinical Intervention Studies on Lycopene

As can be seen in this book, clinical intervention studies with lycopene have been carried in relation to various diseases such as cardiovascular diseases, diabetes, prostate cancer, and so on (see other chapters in this book). Other intervention studies using caroteroids in general have been reported [34,35]. Since our laboratory is the only one to date that reported clinical intervention studies with lycopene in relation to osteoporosis, this section will focus on reviewing the results of our studies.

To establish the role of lycopene in the prevention of risk for osteoporosis in postmenopausal women, Rao et al. carried out four different clinical studies. In the first study, the effects of a lycopene-restricted diet on oxidative stress parameters and bone turnover markers in postmenopausal women were determined [36]. Exclusion criteria were established to avoid the effects of compounding factors with antioxidants. Thus, the following participants were excluded: those using medications that may affect bone metabolism or have antioxidant properties, smokers, and those who could not avoid consuming tomatoes and tomato products. Twenty-three postmenopausal women between the ages of 50 and 60 years provided blood samples and seven-day dietary records. Following a one-month depletion period, during which no lycopene-containing foods were consumed, participants provided another blood sample and dietary records. The blood parameters before and after seven-day restriction diet were analyzed. Results revealed that the overall change in the serum carotenoids in a lycopene-restricted diet was not as high as that seen for participants who consume lycopene, and significant decreases in serum lycopene, α-/β-carotene, and lutein/zeaxanthin were noted. Lycopene restriction also resulted in decreases of all configurations of lycopene (all trans, 5-cis-, and other cis lycopene) (Figure 6.2) as well as those of the antioxidant enzymes SOD and CAT (data not shown). A significant increase in the bone resorption marker NTX was found to accompany these changes when lycopene was restricted (Figure 6.3).

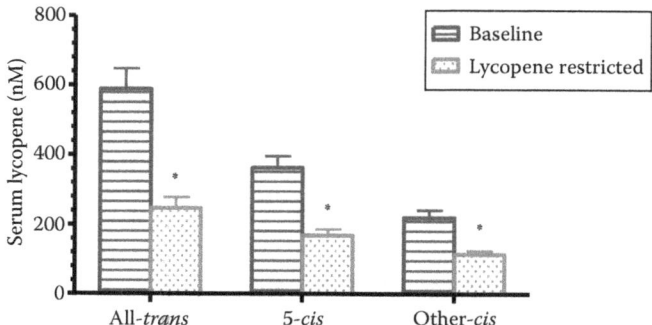

FIGURE 6.2 Decrease in all configurations of lycopene (all trans, 5-cis- and other cis lycopene) in the serum of postmenopausal women after lycopene restriction. (Reproduced from Rao, L. G., and A. V. Rao. "Oxidative Stress and Antioxidants in the Risk of Osteoporosis—Role of the Antioxidants Lycopene and Polyphenols," in: *Topics in Osteoporosis*, ed. Margarita Valdes Flores, 2013. Available at http://www.intechopen.com/articles/show/title/oxidative-stress-and-antioxidants-in-the-risk-of-osteoporosis-role-of-the-antioxidants-lycopene-and-. ISBN: 978-953-51-1066-8. With permission from Intech.)

The dramatic effects of dietary lycopene restriction on increasing the risk for osteoporosis in postmenopausal women form the first evidence proving that lycopene may be beneficial in reducing this risk. An extension to this observation is the possibility of what Brown et al. [37] observed—that a significant increase in the bone resorption marker NTX could lead to a long-term decrease in BMD and increase in fracture risk. A longer restriction period may be even more detrimental to a group of postmenopausal women who are already at high risk for osteoporosis. As a result of this study, a

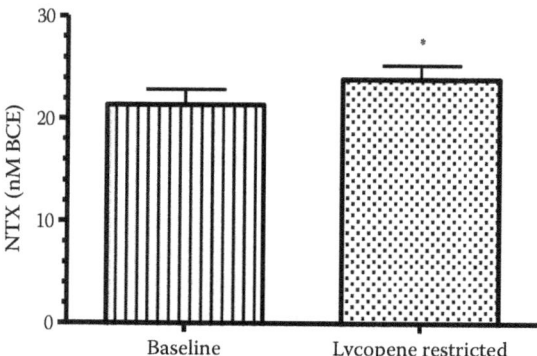

FIGURE 6.3 Increase in the concentrations of the bone resorption marker NTX, in the serum of postmenopausal women after lycopene restriction. (Reproduced from Rao, L. G., and A. V. Rao. "Oxidative Stress and Antioxidants in the Risk of Osteoporosis—Role of the Antioxidants Lycopene and Polyphenols," in: *Topics in Osteoporosis*, ed. Margarita Valdes Flores, 2013. Available at http://www.intechopen.com/articles/show/title/oxidative-stress-and-antioxidants-in-the-risk-of-osteoporosis-role-of-the-antioxidants-lycopene-and-. ISBN: 978-953-51-1066-8. With permission from Intech.)

shorter wash-out period of no lycopene consumption is recommended in clinical trials examining the effects of lycopene on bone health [28].

The second study carried out was a clinical fully randomized controlled intervention study by Mackinnon et al. [38] to directly investigate the effects of lycopene supplementation on decreasing the risk for osteoporosis. Sixty postmenopausal women of 50 to 60 years old were recruited. The exclusion criteria were similar to those used during the first study. Participants were assigned to one of the following four groups (N = 15/group) after a washout period of one month without lycopene consumption: a group consuming (1) regular tomato juice, (2) lycopene-rich tomato juice, (3) tomato lycopene capsules, or (4) placebo capsules, to be consumed twice daily for total lycopene intakes of 30, 70, 30, and 0 mg/day, respectively, for four months. The collected sera were assayed for oxidative stress parameters, total antioxidant capacity, and bone turnover markers. Results showed that the placebo had lower serum lycopene than the lycopene-supplementation for four months and the difference was very significant ($P <$ 0.001). Since the increase in serum lycopene was similar for all three supplements, the results were pooled into a placebo-supplemented and lycopene-supplemented group for further statistical analyses. After the four-month duration, the lycopene-supplemented group had a significant increase in total antioxidant capacity (Figure 6.4), a decrease in the oxidative stress parameter of protein oxidation, as shown by an increase in thiol values (Figure 6.5) and lipid peroxidation, as shown by TBARS (Figure 6.6), which correlated to a decrease in NTX (Figure 6.7); all changes were significantly different from the placebo group. These findings suggest that lycopene obtained in the form of tomato juice or capsules exerted equivalent antioxidant potency in reducing the risk of osteoporosis in postmenopausal women [38].

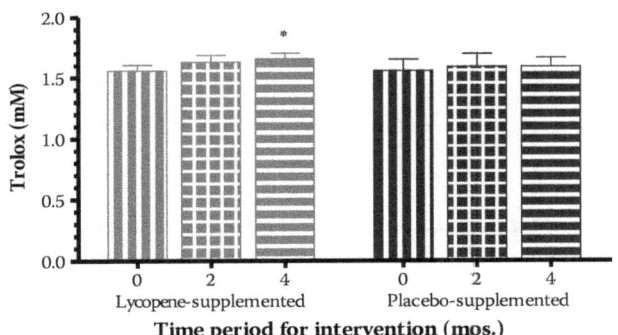

FIGURE 6.4 Increase in the serum total antioxidant capacity of postmenopausal women supplemented with LYCOPENE compared to placebo capsules for 4 months. Values are mean ± SEM. Values compared within supplement group was determined to be statistically significant using repeated-measures ANOVA (*p<0.05). (Reproduced from Rao, L. G., and A. V. Rao. "Oxidative Stress and Antioxidants in the Risk of Osteoporosis—Role of the Antioxidants Lycopene and Polyphenols," in: *Topics in Osteoporosis*, ed. Margarita Valdes Flores, 2013. Available at http://www.intechopen.com/articles/show/title/oxidative-stress-and -antioxidants-in-the-risk-of-osteoporosis-role-of-the-antioxidants-lycopene-and-. ISBN: 978-953-51-1066-8. With permission from Intech.)

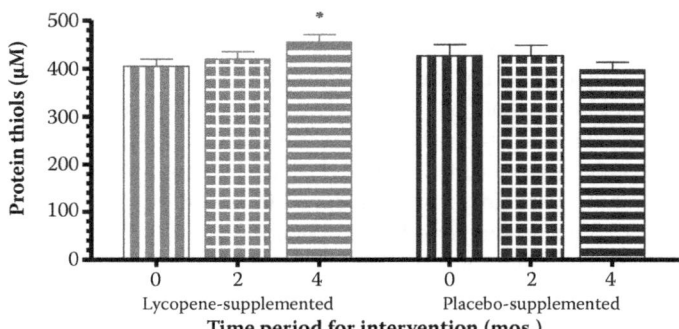

FIGURE 6.5 Increase in the serum concentration of thiol (meaning decreased protein oxidation) in postmenopausal women supplemented with LYCOPENE compared to placebo capsules for a period of 4 months. Values are mean ± SEM. Values compared within supplement group were determined to be statistically significant using repeated-measures ANOVA (*p<0.001). (Reproduced from Rao, L. G., and A. V. Rao. "Oxidative Stress and Antioxidants in the Risk of Osteoporosis—Role of the Antioxidants Lycopene and Polyphenols," in: *Topics in Osteoporosis*, ed. Margarita Valdes Flores, 2013. Available at http://www.intechopen.com/articles/show/title/oxidative-stress-and-antioxidants-in-the -risk-of-osteoporosis-role-of-the-antioxidants-lycopene-and-. ISBN: 978-953-51-1066-8. With permission from Intech.)

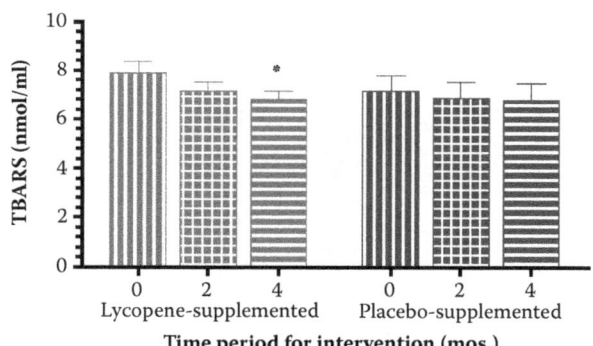

FIGURE 6.6 Decrease in the serum concentration of TBARS or lipid peroxidation in postmenopausal women supplemented with LYCOPENE compared to placebo capsules for a period of 4 months. Values are mean ± SEM. Values compared within supplement group were determined to be statistically significant using repeated-measures ANOVA (*p<0.001). (Reproduced from Rao, L. G., and A. V. Rao. "Oxidative Stress and Antioxidants in the Risk of Osteoporosis—Role of the Antioxidants Lycopene and Polyphenols," in: *Topics in Osteoporosis*, ed. Margarita Valdes Flores, 2013. Available at http://www.intechopen.com/articles/show/title/oxidative-stress-and-anti oxidants-in-the-risk-of-osteoporosis-role-of-the-antioxidants-lycopene-and-. ISBN: 978-953-51-1066-8. With permission from Intech.)

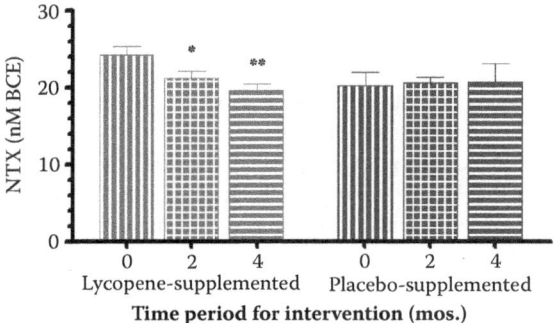

FIGURE 6.7 Decrease in the serum concentration of bone resorption marker NTX in post-menopausal women supplemented with LYCOPENE compared to placebo capsules for a period of 4 months. Values are mean ± SEM. Values compared within supplement group were determined to be statistically significant at 2 and 4 months using repeated-measures ANOVA (*$p < 0.01$ and **$p < 0.001$). (Reproduced from Rao, L. G., and A. V. Rao. "Oxidative Stress and Antioxidants in the Risk of Osteoporosis—Role of the Antioxidants Lycopene and Polyphenols," in: *Topics in Osteoporosis*, ed. Margarita Valdes Flores, 2013. Available at http://www.intechopen.com /articles/show/title/oxidative-stress-and-antioxidants-in-the-risk-of-osteoporosis-role-of-the -antioxidants-lycopene-and-. ISBN: 978-953-51-1066-8. With permission from Intech.)

Mackinnon et al. [38] carried out a third study to compare bone turnover markers, oxidative stress parameters, and serum lycopene data between postmenopausal women who obtained both a low and a high intake of lycopene from their daily food diet with those women who were supplemented with lycopene. The objective of the study was to determine whether the added dose through supplementation contributed to the beneficial effect of lycopene in reducing bone turnover markers when compared to regular intakes typically obtained from the usual daily diet. Here we compare data from participants in our intervention study—in which bone turnover markers, antioxidant capacity, and oxidative stress parameters were significantly affected—to data from participants in our cross-sectional study of the same age who consumed their usual lycopene intake. Results showed that women who consumed lycopene supplements had significantly lower TBARS values than participants who had no lycopene supplementation but who obtained a low intake or high intake of lycopene through their usual daily diets. These differences in TBARS values may be attributed to a significantly higher concentration of serum 5-*cis* in lycopene-supplemented participants compared to participants who obtained lycopene from their usual daily diet. These results suggest that the 5-*cis* isomer has the ability to provide the greatest protection against oxidative stress because of its potent antioxidant capacity, which, at higher concentrations, decreases bone turnover markers. It may also be concluded that supplementation with lycopene may be necessary in spite of a daily intake of lycopene.

A fourth study was carried out to test whether polymorphism of paraoxonase 1 (PON 1) has a role in the effect of lycopene on preventing the risk for osteoporosis in postmenopausal women. Human paraoxonase (PON1) is an HDL-associated, calcium-dependent esterase enzyme that has been shown to hydrolyze hydrogen peroxide, a potent ROS. These mechanisms prevent the accumulation of damaging ROS and decrease oxidative stress in the body.

The results showed that 172T∀A or 584A∀G polymorphisms of PON1 modified the association between lycopene, NTX, and BAP—an interaction that may also moderate the risk of osteoporosis [31].

In another study, Mackinnon et al. [39,40] showed that there was a significant interaction between the PON1 genotype and changes in TBARS ($P < 0.05$), suggesting that supplementation with lycopene resulted in decreased lipid peroxidation, which interacted with the PON1 genotype to decrease bone resorption markers in postmenopausal women. These results provided mechanistic evidence of how intervention with lycopene may act to decrease lipid peroxidation and thus the risk of osteoporosis in postmenopausal women [39,40].

6.3　GENERAL SUMMARY AND CONCLUSION

This review clearly demonstrates that: (1) oxidative stress is caused by reactive oxygen species (ROS); (2) oxidative stress is shown to be associated with the development of osteoporosis; (3) the damaging effects of oxidative stress may be preventable by supplementation with the antioxidant lycopene; (4) these conclusions are supported by results of *in vitro* studies in osteoblasts and osteoclasts, and epidemiological and clinical intervention studies with the antioxidant lycopene; and (5) lycopene may have the potential for use as an alternative or complementary agent with other established drugs approved for the prevention or treatment of osteoporosis in postmenopausal women.

ACKNOWLEDGMENTS

Funding for this research into oxidative stress, antioxidants, and bone health is shared by Genuine Health Ltd (Canada), the H.J. Heinz Co (USA), Millenium Biologix Inc. (Canada), Kagome Co. (Japan), and LycoRed Natural Product Industries, Ltd. (Israel) and matched by the Canadian Institutes of Health Research (CIHR). We sincerely thank the valuable contributions to this research by the following undergraduate summer students, graduate students, postdoctoral fellows, and staff at the Calcium Research Laboratory, Department of Medicine at St Michael's Hospital and the Department of Nutritional Sciences, University of Toronto: Bala Balachandran, Jaclyn Beca, Dawn Snyder, Loren Chan, Honglei Shen, Salva Sadeghi, Ayesha Quireshi, Erin Mackinnon, and Nancy Kang. Their contributions were based on their experimental data, written reports, and published manuscripts. We would also like to thank Alan Logan and R. G. Josse for providing us with their medical expertise, as well as allowing us access to their list of patients. Special thanks go to H. Vandenberghe for carrying out the CTX assay and for her valuable suggestions.

LIST OF ABBREVIATIONS

CTX　　　　crosslinked C-telopeptides of type I collagen
HRT　　　　Hormone Replacement Therapy

NTX	crosslinked N-telopeptides of type I collagen
ROS	reactive oxygen species
TBARS	thiobarbituric acid reactive substances
TEAC	Trolox equivalent antioxidant capacity

REFERENCES

1. Ahmed, S., and M. Elmantaser. 2009. "Secondary osteoporosis." *Endocr Dev* 16:170–90.
2. Gallagher, J., and A. Sai. 2010. "Molecular biology of bone remodelling: Implications for new therapeutic targets for osteoporosis." *Maturitas* 65:301–7.
3. Rao, L. G., and A. V. Rao. 2013. "Oxidative Stress and Antioxidants in the Risk of Osteoporosis—Role of the Antioxidants Lycopene and Polyphenols," in *Topics in Osteoporosis*, ed. Margarita Valdés-Flores, InTech, DOI: 10.5772/54703. Available from: https://www.intechopen.com/books/topics-in-osteoporosis/oxidative-stress-and -antioxidants-in-the-risk-of-osteoporosis-role-of-the-antioxidants-lycopene-and- (available May 15, 2015).
4. Stetzer, E. 2011. "Identifying risk factors for osteoporosis in young women." *IJAHSP* 9:1–8.
5. Semba, R. D., L. Ferrucci, K. Sun, J. Walston, R. Varadhan, J. M. Guralnik, and L. P. Fried. 2007. "Oxidative stress is associated with greater mortality in older women living in the community." *J Am Geriatr Soc* 55:1421–5.
6. Schmidt, G. A., K. E. Horner, D. L. McDanel, M. B. Ross, and K. G. Moores. 2010. "Risks and benefits of long-term bisphosphonate therapy." *Am J Health Syst Pharm* 67:994–1001.
7. Diab, D., and N. Watts. 2012. "Bisphosphonates in the treatment of osteoporosis." *Endocrinol Metab Clin North Am* 41:487–506.
8. Rees, M. 2011. "Management of the menopause: Integrated health-care pathway for the menopausal woman." *Menopause Int* 17:50–4.
9. Pomerleau, J., K. Lock, C. Knai, and M. Mackee. 2005. "Effectiveness of intervention and programmes promoting fruit and vegetable intake." *J Nutr.* 135:2486–2495.
10. Rao, A. V., and L. G. Rao. 2007. "Carotenoids and human health (Review)" *Pharmacol Res* 55:207–16.
11. Tucker, K., H. Hannan, H. Chen, L. Cupples, P. Wilson, and D. Kiel. 1999 "Potassium, magnesium, and fruit and vegetable intakes are associated with greater bone mineral density in elderly men and women." *Am J Clin Nutr* 69:727–36.
12. New, S. A. 2003. "Intake of fruit and vegetables: Implications for bone health." *Proc Nutr Soc* 62:889–99.
13. Lister, C., M. Skinner, and D. Hunter. 2007. "Fruits, vegetables and their phytochemi-cals for bone and joint health." *Curr Top Nutraceut Res* 5:67–82.
14. Mate, J. M., C. Perez-Gomez, and I. Nunez de Castro. 1999. "Antioxidant Enzymes and Human Diseases." *Clin Biochem* 32:595–603.
15. Rao, A. V., M. Ray, and L. Rao. 2006. "Lycopene," in: *Advances in Food and Nutrition Research*, ed. S. L. Taylor, Academic Press Publication, New York, pp. 99–164.
16. Kim, L., A. V. Rao, and L. G. Rao. 2003. "Lycopene II. Effect on osteoblasts: The carotenoid lycopene stimulates cell proliferation and alkaline phosphatase activity of SaOS-2 cells." *J Med Food* 6:79–88.
17. Park, C. K., Y. Ishimi, M. Ohmura, M. Yamaguchi, and S. Ikegami. 1997. "Vitamin A and carotenoids stimulate differentiation of mouse osteoblastic cells." *J Nutr Sci and Vitaminol* 43:281–96.

18. Rao, L. G., A. V. Rao, and E. S. Mackinnon. 2013. "CIS-lycopene prevents and repairs the damaging effects of reactive oxygen species in human osteoblast cells." Asia Pacific International Osteoporosis Foundation (IOF) Regionals Osteporosis Meeting, Hong Kong, December 12–15, 2013.
19. MacKinnon, E. 2010. "The Role of the Carotenoid Lycopene as an Antioxidant to Decrease Osteoporosis Risk in Women: Clinical and in vitro Studies." Dissertation for PhD, Institute of Medical Science. Toronto, Ontario: University of Toronto.
20. Rao, L., N. Krishnadev, K. Banasikowska, and A. V. Rao. 2003. "Lycopene I. Effect on Osteoclasts; Lycopene inhibits basal and parathyroid hormone (PTH)-stimulated osteoclast formation and mineral resorption mediated by reactive oxygen species (ROS) in rat bone marrow cultures." J Med Food 6:69–78.
21. Ishimi, Y., M. Ohmura, X. Wang, M. Yamaguchi, and S. Ikegami. 1999. "Inhibition by carotenoids and retinoic acid of osteoclast-like cell formation induced by bone resorbing agents in vitro." J Clin Biochem Nutr 27:113–22.
22. Serra-Majem, L., B. Roman, and R. Estruch. 2006. "Scientific evidence of interventions using the Mediterranean diet: A systematic review." Nut Rev 64:S27–S47.
23. Hagfors, L., P. Leanderson, L. Skoldstam, J. Andersson, and G. Johansson. 2003. "Antioxidant intake, plasma antioxidants and oxidative stress in a randomized, controlled, parallel, Mediterranean dietary intervention study on patients with rheumatoid arthritis." Nutr J 2:5.
24. Leighton F., A. Cuevas, V. Guasch, D. D. Pérez, P. Strobel, A. San Martín, U. Urzua, M. S. Díez, R. Foncea, O. Castillo, C. Mizón, and M. A. Espinoza. 1999. "Plasma polyphenols and antioxidants, oxidative DNA damage and endothelial function in a diet and wine intervention study in humans." Drugs Exp Clin Res 25:133–41.
25. Puel, C., V. Coxam, and M. Davicco. 2007. "Mediterranean diet and osteoporosis prevention." Med Sci (Paris) 23:756–60.
26. Yang, Z., Z. Zhang, K. Penniston, N. Binkley, and S. Tanumihardjo. 2008. "Serum carotenoid concentrations in postmenopausal women from the United States with and without osteoporosis." Int J Vitam Nutr Res 78:1050–111.
27. Sahni, S., M. Hannan, J. Blumberg, L. Cupples, D. Kiel, and K. Tucker. 2009. "Protective effect of total carotenoid and lycopene intake on the risk of hip fracture: A 17-year follow-up from the Framingham Osteoporosis Study." J Bone Miner Res 24:1086–94.
28. Sugiura, M. 2008. "Bone mineral density in post-menopausal female subjects is associated with serum antioxidant carotenoids." Osteoporosis Int 19:211–19.
29. Wattanapenpaiboon, N., W. Lukito, M. Wahlqvist, and B. Strauss. 2003. "Dietary carotenoid intake as a predictor of bone mineral density." Asia Pac J Clin Nutr 12:467–73.
30. Rao, L., E. Mackinnon, R. Josse, T. Murray, A. Strauss, and A. V. Rao. 2007. "Lycopene consumption decreases oxidative stress and bone resorption markers in postmenopausal women." Osteoporosis Int 18:109–15.
31. Liang, H., F. Yu, Z. Tong, and W. Zeng. 2012. "Lycopene effects on serum mineral elements and bone strength in rats." Molecules 17(6):7093–102.
32. Iimura, Y., U. Agata, S. Takeda, Y. Kobayashi, S. Yoshida, I. Ezawa, and N. Omi. 2015. "The protective effect of lycopene intake on bone loss in ovariectomized rats." J Bone Miner Metab 33:270–78.
33. Ardawi, M. S., M. H. Badawoud, S. M. Hassan, N. M. AlNosani, M. H. Qari, and S. A. Mousa. 2016. "Lycopene treatment against loss of bone mass, microarchitecture and strength in relation to regulatory mechanisms in a postmenopausal osteoporosis model." Bone 83:127–40.
34. Zang, Z. Q., W.-T. Cao, J. Liu, Z. Q. Zhang, W. T. Cao, J. Liu, Y. Cao, Y. X. Su, Y. M. Chen. 2016. "Greater serum carotenoid concentration associated with higher bone mineral density in Chinese adults." Osteoporosis Int 27:1593–1601.

35. Hayhoe, R. P. G., M. A. H. Lentjes, A. A. Mulligan, R. N. Luben, K.-T. Khaw, and A. A. Welch. 2017. "Carotenoid dietary intakes and plasma concentrations are associated with heel bone ultrasound attenuation and osteoporotic fracture risk in the European Prospective Investigation into Cancer and nutrition (EPIC)-Norfolk cohort." *Br J Nutr* 117:1439–53.

36. Mackinnon, E., A. V. Rao, and L. Rao. 2011. "Dietary restriction of lycopene for a period of one month resulted in significantly increased biomarkers of oxidative stress and bone resorption in postmenopausal women." *J Nutr Health Aging* 15:133–8.

37. Brown, J., C. Albert, B. Nassar, J. D. Adachi, D. Cole, K. S. Davison, K. C. Dooley, A. Don-Wauchope, P. Douville, D. A. Hanley, S. A. Jamal, R. Josse, S. Kaiser, J. Krahn, R. Krause, R. Kremer, R. Lepage, E. Letendre, S. Morin, D. S. Ooi, A. Papaioaonnou, and L. G. Ste-Marie. 2009. "Bone turnover markers in the management of postmenopausal osteoporosis." *Clin Biochem* 42:929–42.

38. Mackinnon, E., A. V. Rao, R. Josse, and L. Rao. 2011. "Supplementation with the antioxidant lycopene significantly decreases oxidative stress parameters and the bone resorption marker N-telopeptide of type I collagen in postmenopausal women." *Osteoporosis Int* 22:1091–101.

39. Mackinnon, E., A. El-Sohemy, A. V. Rao, and L. Rao. 2010. "Paraoxonase 1 polymorphisms 172T->A and 584A->G modify the association between serum concentrations of the antioxidant lycopene and bone turnover markers and oxidative stress parameters in women 25–70 years of age." *J Nutrigenet Nutrigenomics* 3:1–8.

40. Rao, L., E. Mackinnon, A. El-Sohemy, and A. V. Rao. 2010. "Postmenopausal Women with PON1 172TT Genotype Respond to Lycopene Intervention With a Decrease in Oxidative Stress Parameters and Bone Resorption Marker NTx." Annual Meeting of the American Society for Bone and Mineral Research (ASBMR); October 13–16, 2010; Toronto, Ontario.

7 Lycopene, Tomatoes, and Male Infertility

Yu Yamamoto, Teruaki Iwamoto, and Ikuo Sato

CONTENTS

7.1   INTRODUCTION

7.1.1   Infertility

Infertility is a condition of the reproductive system that prevents the conception of children. According to the International Committee for Monitoring Assisted Reproductive Technology, World Health Organization (WHO), infertility is a disease of the reproductive system defined as the failure to achieve clinical pregnancy after ≥ 12 months of regular unprotected sexual intercourse [1].

Many couples worldwide experience infertility and seek help to achieve pregnancy. The WHO estimates that 60–80 million couples worldwide are currently infertile [2], although the global prevalence rates of infertility are difficult to determine. Because one in eight couples is estimated to have problems in achieving or sustaining a pregnancy, infertility can be considered one of the biggest health problems worldwide. It is reported that most infertile patients live in developing countries. According to one study published at the end of 2012 by the WHO, one in every four couples in developing countries had been found to be affected by infertility [3]. Even in Japan, a developed country, the number of patients treated for infertility exceeds 50,000, and that of potential infertile patients is considered to be quite large.

7.1.2   Male Infertility

Male infertility refers to the inability of a male individual to cause pregnancy in a fertile female individual. In humans, male infertility accounts for 40–50% of all cases of infertility [4,5] (Figure 7.1) and affects ~7% of all men [6].

"Male factor" infertility is seen as an alteration in the sperm concentration and/or motility and/or morphology in at least one of two analyzed semen samples collected one and four weeks apart [7]. The WHO has revised the lower reference limits for semen analysis, which are widely used as a basis for the standard diagnosis of male infertility. The normal semen analysis values (WHO, 5th edition) are described in Table 7.1.

The causes and percentage of distribution of male infertility are shown in Figure 7.2. The chart was created in a study on 12,945 male infertile patients attending the Institute of Reproductive Medicine of the University of Münster [8]. The most common cause is "Infertility of known (possible) cause," which includes varicocele,

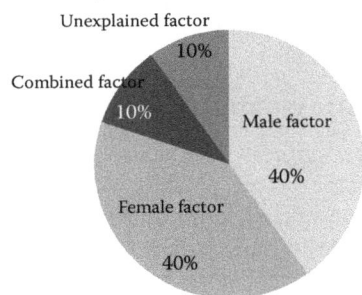

FIGURE 7.1 Causes of infertility among men and women.

TABLE 7.1
WHO Guidelines for Normal Semen Parameters

Semen Analysis Parameters	Normal Values
Volume	≥ 1.5 mL
pH	≥ 7.2
Sperm concentration	≥ 15,000,000/mL
Total motility	≥ 40%
Progressive motility	≥ 32%
Morphology	≥ 4% normal forms (strict criteria)
Vitality	≥ 58% live sperm
White blood cells	< 1,000,000/mL

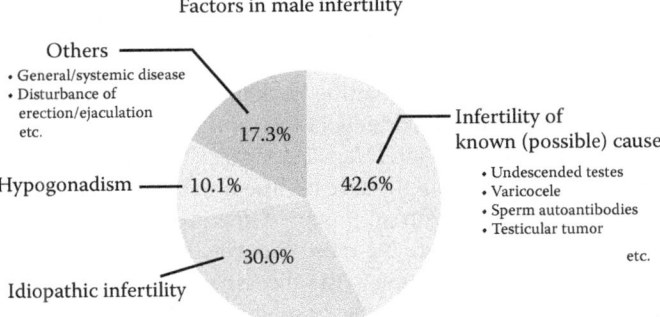

FIGURE 7.2 Factors contributing to male infertility.

maldescended testes, sperm autoantibodies, and testicular tumor. In about 30% of cases, no factor associated with male infertility was found (idiopathic male infertility). Men with idiopathic infertility present with no history of diseases affecting fertility and have normal findings on physical examination as well as in endocrine, genetic, and biochemical laboratory tests; however, semen analysis might reveal pathological findings. Although the entire mechanism is unclear, idiopathic male infertility is assumed to be caused by several factors, including alcohol consumption, cigarette smoking, and environmental pollutants.

7.1.3 TREATMENT FOR MALE INFERTILITY

For male infertile patients, surgical treatment or medical treatment is performed depending on the cause. For example, patients with varicocele and non-obstructive or obstructive azoospermia especially often undergo surgical operations as a therapy for male infertility. For male infertile patients having problems with erectile dysfunction, medicines, such as sildenafil (Viagra), ejaculation, or stimulation of certain nerves using electrodes may help. Penile electroejaculation stimulates the nerves involved in ejaculation.

For azoospermic patients with unknown hypogonadotropic hypogonadism, gonadotropin and recombinant human follicle stimulating hormone (r-hFSH) self-injection are useful treatments aiming to stimulate the testes to produce testosterone. This treatment is very effective when the sperm appears in the ejaculated semen, and natural pregnancy can be expected [9].

Conversely, many male infertile patients have no male infertility-associated factor (idiopathic male infertility). Some couples also take measures, such as intrauterine insemination, *in vitro* fertilization, and intracytoplasmic sperm injection, to conceive children. Meanwhile, in cases of idiopathic male infertility, many urologists prescribe oral antioxidant therapies for men with clinical subfertility [10]. The rationale for recommending oral antioxidant therapies is based on the premise that seminal oxidative stress is due, in part, to a deficiency in seminal antioxidants and the lack of serious adverse effects rather than other kinds of therapies [11].

7.2 MALE INFERTILITY AND REACTIVE OXYGEN SPECIES (ROS)

Although the reason why the abovementioned factors cause male infertility has not been fully clarified, a recent report identified oxidative stress as a likely cause [12]. It was shown that the levels of antioxidants in the seminal plasma of fertile men were significantly lower than the levels in infertile controls [13]. In addition, it was shown that up to 40% of infertile men might have high seminal ROS levels [14]. High levels of ROS have been associated with an impaired sperm motility, concentration, and morphology. These parameters are the most important predictors of the potential to produce viable sperm [15]. In addition to the abovementioned reports, several studies have identified oxidative stress as a cause of male infertility.

However, ROS itself does not necessarily cause male infertility. In the spermatozoa, leukocytes, neutrophils, macrophages, and immature cells are considered the main endogenous sources of ROS, and appropriate amounts of ROS are known to be useful for the maintenance of sperm function. Several studies have indicated that the spermatozoa need small amounts of ROS to acquire the ability for capacitation, hyperactivation, motility, acrosome reaction, and fertilization [16,17]. Only when ROS occurs at high concentrations in the spermatozoa and overcomes the antioxidant defense systems do pathological defects occur.

In this section, we describe the mechanism by which high concentrations of ROS cause male infertility and the causes of ROS production.

7.2.1 Mechanisms by Which ROS Causes Male Infertility

7.2.1.1 Lipid Peroxidation

Lipids in the surface of human spermatozoa are the major substances responsible for the fluidity of membrane lipid bilayers, which alter the composition of the plasma membranes of sperm cells from their epididymal maturation to their capacitation in the female reproductive tract [18]. As lipids in the surface of human spermatozoa are oxidized by ROS, this causes sperms to lose their function by decreasing their fluidity and integrity.

As a characteristic feature of the lipid composition of human spermatozoa membranes, the plasma membrane contains high levels of lipids in the form of polyunsaturated fatty acids (PUFAs), which make it more prone to autoxidation when exposed to ROS. Because the plasma membrane contains high levels of PUFAs, the human spermatozoa are highly susceptible to oxidative stress. Lipid peroxidation can lead to loss of membrane fluidity and integrity. The membrane lipid composition has been related to the specific functions of each compound, as they promote the creation of microdomains with different fluidity, fusogenicity, and permeability characteristics required for reaching and fusing with the oocyte [19].

Docosahexaenoic acid (DHA), a long-chain fatty acid with 22 carbon chains and six *cis* double bonds, is the most common PUFA present. DHA accounts for ~50% of the fatty acids in the human spermatozoa. Owing to the high number of double bonds, DHA is particularly susceptible to peroxidative breakdown. As PUFAs, such as DHA, are oxidized by ROS, they cause sperms to lose their function by decreasing their fluidity and integrity. The acyl chains of DHA bound to the phospholipids of sperm membranes are particularly susceptible to ROS attack, leading to the formation of 4-hydroxynonenal and malondialdehyde (MDA). Both of these toxic molecules can then cause protein alterations and DNA mutations [20]. Hydroxynonenals are hydrophilic and can cause severe cell dysfunctions at both genomic and proteomic levels [21]. As MDA is produced as an end-product of lipid peroxidation, semen MDA concentrations can be an indicator of the damage caused to the lipid membranes in human sperm.

7.2.1.2 DNA Damage

Many studies have investigated the role of sperm DNA damage in male-factor infertility. There is evidence that the sperm of infertile men contain more DNA damage than those of fertile men and that sperm DNA damage may have a negative effect on fertility potential [22–24]. High levels of sperm DNA damage often correlate with poor semen parameters, such as reduced count and motility, or abnormal morphology [25,26]. Although the mechanism by which sperm DNA damage affects reproductive outcomes is far from being completely understood, we describe here the current knowledge about the mechanism of sperm DNA damage by ROS.

Although the cause of sperm DNA damage is believed to be multifactorial, e.g., defects during spermatogenesis and abortive apoptosis, the main cause of sperm DNA damage is believed to be oxidative in nature [27]. A positive correlation was shown between sperm DNA fragmentation and ROS [28]. Both mitochondrial and sperm nuclear DNA are potential targets of attack by ROS; sperm nuclear DNA damage induced by ROS is more well known.

In the nuclear DNA of sperm, the chromatin has a highly condensed and organized structure, which protects against oxidative damages, making sperm DNA very resistant to damage [29]. However, when the compaction is poor and chromatin protamination is incomplete, sperm DNA is more vulnerable to ROS. High ROS concentrations in infertile men have been associated with poor chromatin packing. After being attacked by ROS, the nuclear DNA shows fragmentation, denaturation, and accumulation of highly mutagenic oxidized adducts, such as 8-hydroxy-2-deoxyguanosine (8-OHdG). 8-OHdG is used as a ubiquitous marker of oxidative stress, and estimating the levels of 8-OHdG in the seminal plasma of infertile patients is an efficient and reliable method of analyzing oxidative damage to the sperm nucleus [30].

7.2.1.3 Apoptosis

Apoptosis plays an important role in male reproductive function. During early development, apoptosis is necessary in ensuring that there is an optimal ratio of Sertoli cells to germ cells, and this ratio is very important in spermatogenesis. In adult life, apoptosis ensures the selective deletion of differentiating germ cells that have been damaged for a wide variety of reasons [31].

However, high levels of ROS cause excessive sperm apoptosis, which adversely affects male infertility. Moustafa et al. determined that infertile patients had high ROS levels in their seminal plasma and a higher percentage of apoptosis than normal healthy donors [32].

Apoptosis of the spermatozoa has been reported to have effects on semen parameters. Fathi et al. conducted a meta-analysis of nine articles. They observed that increased apoptosis intensity decreases sperm motility and affects sperm concentration [33].

The action of the signaling molecule cytochrome c released from the mitochondria is known as one of the mechanisms by which ROS causes sperm apoptosis. ROS serves as a stimulus that activates the mitochondria to release cytochrome c. Cytochrome c initiates a cascade of events that involve several caspases, including caspase-3 and caspase-9, which lead to sperm apoptosis [34].

7.2.2 EXOGENOUS SOURCES OF ROS

7.2.2.1 Alcohol Consumption

Alcohol consumption is known to promote ROS production and interferes with the antioxidant defense mechanism of the body, particularly in the liver. Alcohol consumption has also been found to interfere with sperm function. In the alcohol metabolism pathway, NADH and acetaldehyde are produced. NADH increases the activity of the respiratory chain in the mitochondria, and acetaldehyde reacts with lipids and proteins to produce ROS in human sperm [35,36].

Therefore, excessive alcohol consumption is associated with a decreased percentage of normal spermatozoa in patients with asthenozoospermia [36]. It has been observed that alcohol consumption increases the abnormalities in the nucleus and plasma membrane of spermatozoa [37,38]. Talebi et al. found that ethanol consumption increases the percentage of sperm cells with chromatin abnormalities [39].

7.2.2.2 Cigarette Smoking

A cigarette contains > 4700 chemical compounds, including alkaloids, nitrosamines, and inorganic molecules. Some of these chemicals are known to cause an imbalance between ROS and antioxidants in the semen [40]. In addition, many researches have reported that cigarette smoking can decrease the level of antioxidants (e.g., vitamins E and C) and increase the ROS level in the seminal plasma [41–44], decreasing the antioxidant ability of the human sperm.

According to the above reports, it is observed that smoking causes deteriorations of sperm function. Cigarette smoking is observed to decrease sperm motility and sperm count [45,46]. Kiziler et al. showed that smokers have increased ROS production with an accompanying decrease in sperm motility [47]. The number of cigarettes smoked per day, number of years smoked, and log-transformed cotinine levels were found to be negatively associated with semen quality, including density, total count, and motility [48].

7.2.2.3 Environmental Pollutants

Many environmental pollutants may be causes of male infertility. For example, chemicals such as dioxin and dichlorodiphenyltrichloroethane, phthalates, and polychlorinated biphenyls are known to interfere with spermatogenesis [49]. Moreover, ingestion of heavy metals can also cause male infertility. A study demonstrated that workers who were regularly exposed to toxins in the form of metals, such as cadmium, chromium, lead, manganese, and mercury, were more likely to have decreased sperm quality, count, volume, and density [50].

Conversely, the mechanism of the effect of these environmental pollutants on male infertility has not been elucidated. Phthalates and nonylphenol have been reported as environmental pollutants causing production of ROS. Phthalates are widely used man-made chemicals that are released to the environment, and their effect on male infertility has been studied in great detail [51]. Phthalates are reported to be contributing factors associated with deterioration in semen quality, and these adverse effects might be mediated by ROS, lipid peroxidation, and mitochondrial dysfunction [52]. Moreover, oral administration of phthalate esters has been found to increase ROS production and reduce the level of antioxidants in the testes, finally causing disturbances in the spermatogenesis [53]. Nonylphenol is an important environmental toxicant and potential endocrine-disrupting chemical. 4-Nonylphenol is found to increase the ROS level in sperms, and a negative correlation was observed between sperm motility and its corresponding ROS level *in vitro* [54]. Moreover, nonylphenol can induce oxidative stress by generating ROSs, such as hydrogen peroxide and superoxide anion, resulting in great damage to the male reproductive system, including altered steroidogenesis, disturbed testicular structure, and decreased sperm number in the epididymis [55].

7.2.2.4 Radiation

Although the number is small, there are reports stating that radiation induces male infertility. Agarwal et al. divided 361 men undergoing an infertility evaluation into four groups according to their active use of a cell phone and measured their sperm parameters. The authors showed that the use of cell phones decreases the semen quality in men by decreasing the sperm count, motility, viability, and normal morphology [56]. Aitken et al. reported that irradiation with radiofrequency electromagnetic radiation induces significant damage to the mitochondrial genome and the nuclear beta-globin locus in rats [57].

Although the mechanism by which radiation causes male infertility has not been clarified, *in vitro* studies have demonstrated that electromagnetic radiation induces ROS production and DNA damage in human spermatozoa [58].

7.2.3 ENDOGENOUS SOURCES OF ROS

7.2.3.1 Varicocele

A varicocele is a dilatation of the pampiniform venous plexus and the internal spermatic vein, which is known to be related to male infertility. Shiraishi et al. reported that the level of seminal ROS is associated with the grade of varicocele: the higher the grade of varicocele, the greater the level of ROS detected [59].

Although it has already been suggested that ROS in varicoceles affects semen parameters, the etiology of elevated oxidative stress levels remains unclear [60]. It was observed that the sperm cells of patients with varicocele have high cytoplasmic droplets producing high levels of ROS [61]. Several studies showed that oxidative stress in men with varicocele could be the result of a decreased total antioxidant capacity [62,63]. Other hypotheses suggested were that ROS in varicocele may be the result of the altered testicular microenvironment and hemodynamics. These alterations stimulate various compensatory mechanisms for maintaining spermatogenesis in patients with varicocele, which cause the upregulation or downregulation of various molecular mechanisms and pathways involved in the generation of free radicals [64].

7.2.3.2 Obesity

The incidence of obesity in male infertile patients has been reported to be three times higher than usual, and obesity has been suggested to induce male infertility [65]. Various causes have been proposed as mechanisms of obesity-induced male infertility, including both hormonal and physical changes. ROS induced by obesity is also proposed to be a cause of male infertility.

Male mice fed with a high-fat diet showed elevated levels of intracellular ROS in the sperm, increased DNA damage, decreased percentage of motile spermatozoa, and a lower number of sperm bound to each oocyte, compared with control mice [66]. It was demonstrated that ROS overproduction and abnormal hormonal regulation in obese men lead to decreased semen quality [67]. Another report suggested that dysregulation of adipocytokines and ROS generation are the causes of oxidative stress in patients with obesity [68].

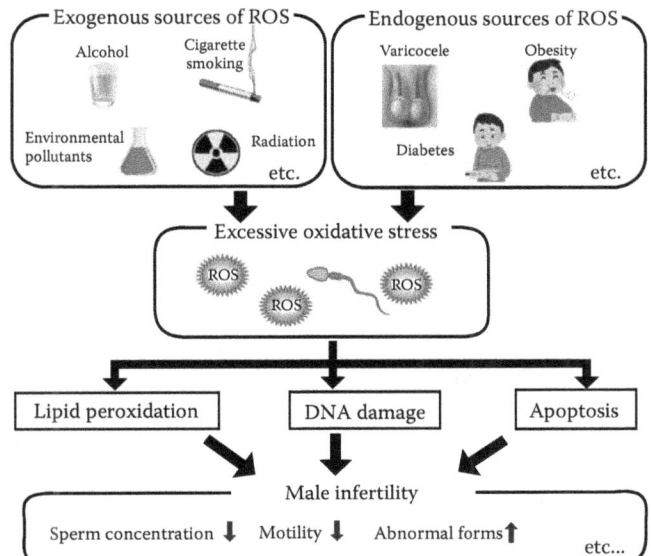

FIGURE 7.3 Mechanisms by which ROS causes male infertility.

7.2.3.3 Diabetes

Several relationships between diabetes and male infertility have been reported, and male infertility is recognized as a complication of diabetes [69]. In patients with diabetes, increased levels of ROS and impairment of the antioxidant defense capacity are observed [70].

It has been shown that there are connections between diabetes and fragmentation of nuclear DNA, deletions of mitochondrial DNA, and oxidative imbalance [71]. Moreover, some research studies demonstrated that the experimental induction of diabetes in animal models induced an impairment of testicular function, progressively leading to decreased fertility [72,73].

As one of the mechanisms of male infertility caused by diabetes, it is suggested that oxidative stress is increased in patients with diabetes mellitus owing to the overproduction of ROS and decreased efficiency of antioxidant defenses [74] through an increased production of advanced glycated end-products and mitochondrial damage [75,76].

In addition, there are reports stating that aging, excessive exercise, and mental stress may cause excessive production of ROS and cause male infertility [77]. Figure 7.3 shows the mechanism by which ROS causes male infertility.

7.3 ANTIOXIDANT TREATMENT

As described in the previous section, an excess production of ROS in the human semen may be one possible cause of male infertility. Therefore, antioxidant treatment is expected to prevent and improve male infertility. The following is a summary of information on frequently reported antioxidants.

7.3.1 Vitamins

The antioxidants α-tocopherol (vitamin E) and ascorbic acid (vitamin C) are potent scavengers of ROS. Many studies have investigated the effects of these vitamins on improving sperm parameters, and some sort of beneficial effects were observed in most of the studies. Two randomized trials showed that vitamin E significantly increased sperm motility and function [78,79]. Moreover, Showell et al. showed that the consumption of vitamin E was associated with a statistically significant increase in the number of live births compared with placebo [80]. Another research shows that the consumption of vitamin C resulted in improvements in sperm quality, and there were statistically significant relationships between serum and seminal plasma vitamin C levels and sperm quality [81]. An antioxidant vitamin intake period of more than three months is recommended because the development of mature sperm from the spermatogonia occurs in $72 \pm$ four days [12]. Further, based on these reports, more than ~1 g/d of vitamin C or 600 mg/d of vitamin E is considered appropriate for intake.

Conversely, there are also reports that found little or no effect of vitamins E and C on the semen parameters. For example, both Kessopoulou et al. [82] and Moilanen and Hovatta [83] reported that the consumption of vitamin E caused no changes in the semen parameters of male infertile patients and healthy volunteers. Rolf et al. found that the consumption of vitamins C and E had no effect on motility, viability, concentration, or morphology in patients with asthenozoospermia [84]. More detailed research is required to clarify the effect of vitamins on male infertility.

7.3.2 Carnitines

Carnitine is a nonprotein amino acid that is known to have an important regulatory role in the mitochondrial transport of long-chain free fatty acids. Many *in vitro* and animal studies have reported that carnitine is a free radical scavenger that protects antioxidant enzymes from oxidative damages [85,86].

The effect of carnitine on male infertility has been observed in several human tests. Some of them have indicated that carnitine has beneficial effects on sperm motility and its concentration [87–89]. Moreover, a systematic review reported that the consumption of carnitines was associated with a significant increase in total sperm motility, forward sperm motility, and atypical sperm forms compared to controls [90].

Meanwhile, what form of carnitine is an active ingredient and what is its mechanism of improving sperm parameters remain unclear. There are some differences in the outcomes depending on the form of carnitine used in each trial (L-carnitine, L-acetyl carnitine, and L-carnitine plus L-acetyl carnitine) [87–89]. Furthermore, Lenzi et al. showed that carnitine levels in the seminal fluid did not change after carnitine consumption [91].

7.3.3 Zinc

Zinc is a cofactor of many metalloproteins for DNA transcription and protein synthesis. It also plays antioxidant roles; zinc deficiency is commonly related to an increase in oxidants, cellular damages, and modulation of antioxidant defenses [92]. In addition, zinc is necessary for the maintenance of spermatogenesis and optimal function of the testes, prostate, and epididymis [93]. From these backgrounds, the consumption of zinc may have the effect of preventing or improving male infertility.

Although there are still few examples of evaluating the influence of zinc on male infertility, it was show that zinc concentrations in the seminal fluid were lower in infertile men [94]. Omu et al. reported that the consumption of zinc improves sperm parameters in men with asthenozoospermia probably through its membrane-stabilizing effect as an antioxidant [95]. However, Wong et al. found that there were no significant improvements in any of the semen parameters, most notably concentration, as determined according to strict criteria in subfertile men [96]. More investigation results are awaited to confirm the effects of zinc on male infertility.

7.3.4 OTHERS

In addition to the abovementioned antioxidant substances with effects on male infertility, selenium, N-acetyl-L-cysteine, coenzyme Q10, and other antioxidants have also been reported [97]. Furthermore, there are some reports on the effects of antioxidant mixtures on semen parameters. However, there are few reports on each substance, and more reports are needed to make a definitive conclusion.

7.4 LYCOPENE TREATMENT

7.4.1 LYCOPENE

Lycopene is a red pigment found in fruits and vegetables, including tomatoes, watermelon, and apricots, and is reportedly one of the most efficient singlet oxygen quenchers and potent peroxyl radical scavengers [98].

Many reports have indicated that lycopene has a beneficial role in preventing chronic diseases, such as cardiovascular disease, atherosclerosis, and cancer [99]. Humans cannot synthesize lycopene and need to consume vegetables and fruits that contain this compound [100]. Lycopene absorbed from the intestines is carried by the bloodstream included in lipoprotein particles and is distributed to various tissues.

Lycopene is reportedly highly concentrated in the male testis. Stahl et al. measured the levels of lycopene and beta-carotene in serum and seven human tissues and found that beta-carotene was the major carotenoid in the liver, adrenal gland, kidney, ovary, and fat, whereas lycopene was the predominant carotenoid in the testes (Table 7.2) [101]. However, the mechanism by which lycopene accumulates at high concentrations in the testes and the role of accumulated lycopene are unknown.

Recent studies have revealed that there is a relationship between semen lycopene level and male infertility. Paren and Naz showed that the concentration of lycopene in the seminal plasma was significantly lower in infertile men (Table 7.3) [102]. With lower levels of antioxidants in the seminal plasma, there will be more free radicals available to cause oxidative damage, resulting in abnormal spermatozoa that cause infertility.

TABLE 7.2
Lycopene Levels in Each Organ

Organ	Lycopene Levels (nmol/g)
Serum	0.29
Liver	1.28
Kidney	0.15
Fat	0.20
Testes	4.34
Ovary	0.25

TABLE 7.3
Lycopene Levels in the Seminal Plasma of Fertile and Infertile Men

	Fertile (n = 22)	Infertile (n = 15)	p Value by ANOVA
Seminal plasma Lycopene (ng/mL)	46.1 ± 33.3	18.8 ± 9.0	0.012

Goyal et al. [103] proved that lycopene concentration in the seminal plasma increases with oral supplementation of lycopene. Therefore, it can be postulated that the intake of lycopene will offer protection from ROS in the seminal plasma and decrease oxidative stress, which is one of the main causes of idiopathic male-factor infertility.

7.4.2 EFFECTS OF LYCOPENE ON SPERM PARAMETERS

From the abovementioned backgrounds, lycopene is expected to improve male infertility by enhancing the antioxidant capacity of the sperm. This hypothesis has prompted a few trials. Table 7.4 shows the summary of the major reports that examined the effects of lycopene consumption on sperm parameters in animal and clinical trials.

7.4.2.1 Animal Trials

Hekimoglu et al. investigated the therapeutic effects of lycopene (4 mg/kg daily, 30 days) on ischemia/reperfusion injury rat. Treatment with lycopene increased sperm motility in both testes and decreased the rate of abnormal sperms in the ipsilateral testes to the sham level, but did not increase sperm concentration in both testes [104].

Mangiagalli et al. investigated the effects of drinking water including 0.1 and 0.5 g/L of lycopene for eight weeks ad libitum on rats. Their data showed that the supplementation of lycopene (0.5 g/L) resulted in a significantly greater volume of ejaculate and total number of sperms but did not affect the sperm concentration [105].

Mangiagalli et al. also showed the therapeutic effects of drinking water supplemented with lycopene (0.5 g/L) for 17 weeks on the semen quality, and therefore

TABLE 7.4
Major Reports that Examined the Effects of Lycopene on Sperm Parameters in Animal and Clinical Trials

Subjects	Dose of Lycopene (Daily)	Duration	Main Results	References
Animal Trials				
42 Rats (sham group, testicular ischemia/reperfusion [IR] injury group, and IR with lycopene group)	4 mg/kg	30 days	Treatment with lycopene increased the motility and decreased the rate of abnormal sperm; however, it did not affect the sperm concentration in bilateral testes.	[104]
18 Rabbits (control group, high or low lycopene group)	0.1 or 0.5g/L (in drinking water)	8 weeks	Lycopene (0.5 g/L) supplementation resulted in a significantly greater volume of ejaculate and total number of sperm, while it did not affects sperm concentration.	[105]
25 Broiler (control group, lycopene group)	0.5 g/L (in drinking water)	17 weeks	Semen production and viability were increased by lycopene supplementation.	[106]
24 Rats (control group, lycopene group, LPS injection group, lycopene + LPS injection group)	4 mg/kg	7 days	The group pretreated with lycopene showed normal sperm parameters compared with the LPS injection group.	[107]
Clinical Trials (Noncontrolled Test)				
30 Infertile men	4 mg/day	3 months	Statistically significant improvement in sperm concentration and motility.	[108]
50 Infertile men	8 mg/day	Until outcome is achieved (at the longest 12, months)	Improvement of sperm count and functional sperm concentration in 70% and 60%, respectively; sperm motility and sperm motility index improved in 54% and 46%, whereas 38% showed improvement in sperm morphology.	[109]

fertility. Their data showed that semen production and viability improved with lyco-pene supplementation. The fertility rate curve of the L group displayed a positive trend [106].

Aly et al. investigated the potential toxicity of lipopolysaccharide (LPS) on the mitochondrial fraction of rat testis and the possible protective efficacy of lycopene (4 mg/kg daily) 24 h before LPS treatment (0.1 mg/kg/day for seven days i.p.). The group pretreated with lycopene showed normal sperm parameters compared with the LPS injection group [107].

Although the number of animal studies that verified the effect of lycopene on male infertility is small, improvement in the sperm quality was recognized in all reports. However, in each report, the animal models and test conditions are different, and there has been no conclusion on the effect of lycopene on male infertility.

7.4.2.2 Human Trials

Gupta and Kumar reported on 30 men with idiopathic nonobstructive oligo-/astheno-/teratozoospermia who were administered 2 mg lycopene twice a day for three months. After administration, the sperm concentration and motility showed a statistically significant improvement [108].

Mohanty et al. reported on 50 patients showing oligoasthenospermia who were administered 8 mg lycopene daily until their sperm analysis results improved to the optimal level or until pregnancy was achieved at the longest timing of 12 months of follow-up. The results showed a 36% pregnancy rate with improvements in the sperm count and functional sperm concentration in 70% and 60% cases, respectively; sperm motility and sperm motility index improved in 54% and 46% cases, respectively, whereas the sperm morphology improved in 38% of the cases [109].

Two human trials analyzed in this review showed that 4–8 mg of daily lycopene supplementation for 3–12 months is sufficient to treat male infertility, improving sperm concentration and motility. However, there were only two non-controlled tests. To elucidate the effect of lycopene on male infertility, a randomized controlled trial needs to be conducted on a larger scale.

7.5 LYCOPENE MECHANISM

As mentioned above, the number of tests that examined the effect of lycopene on male infertility is small. Among those reports, we summarized those that infer the mechanism of lycopene.

7.5.1 EFFECT OF LYCOPENE ON LIPID PEROXIDATION

Semen MDA concentrations can be an indicator of the damage caused to the lipid membranes in the human sperm. Two studies have reported a decrease in the lipid peroxidation after lycopene was administered to the subjects. Aly et al. also showed that pretreatment with lycopene suppresses the LPS-induced MDA and increases the semen levels of antioxidant markers in rats [107]. Sarkar et al. conducted a study in which 45 patients and 30 healthy controls were given lycopene from various sources (tomato, synthetic, or placebo) as part of their diet for ten weeks. The study revealed that the MDA concentration decreased from the supplementation of various forms of lycopene [110].

Both these studies showed that lycopene is effective in reducing lipid peroxidation, therefore reducing the damages caused by oxidative stress.

7.5.2 EFFECT OF LYCOPENE ON DNA DAMAGE

DNA damage is expressed as the percentage of DNA fragmentation in the sperm. Two studies specifically measured this outcome. Zini et al. observed that the reincubation of washed sperm suspensions with 5 µM/L lycopene prevented hydrogen peroxide-induced DNA fragmentation. Their data suggested that preincubation of spermatozoa with lycopene offers protection against oxidative DNA damages *in vitro* [111]. Rosato et al. showed that the presence of lycopene in the extender improved the survival of turkey spermatozoa after liquid storage and protected DNA integrity against cryodamages [112].

Although the number of reports is small, these studies imply that lycopene can aid in the reduction of DNA damages in the spermatozoa.

7.5.3 EFFECT OF LYCOPENE ON APOPTOSIS

Sperm apoptosis is expressed as a caspase activity. There has been no report evaluating the effect of lycopene on apoptosis-related markers in the sperm. There are some reports stating that lycopene consumption increases the sperm concentration and sperm count [108,109]. It is also possible that lycopene inhibits apoptosis through ROS.

The effects of lycopene on lipid peroxidation and DNA damage are believed to be because of its antioxidant action as a singlet oxygen quencher and peroxyl radical scavenger. Meanwhile, lycopene is also known to induce phase II detoxifying/antioxidant enzymes, as found in both *in vitro* and *in vivo* studies [113–115]. Therefore, the possibility that lycopene promotes suppression of lipid peroxidation and DNA damage by promoting the production of antioxidant substances in the sperm is undeniable.

7.6 EFFECTS OF TOMATO JUICE ON MALE INFERTILITY

To evaluate the effect of consumption of foods containing lycopene, we conducted an intervention study to clarify the beneficial effects of tomato juice, which is rich in lycopene, on male fertility.

7.6.1 SUBJECTS

We recruited male infertile patients with poor sperm concentrations ($< 20 \times 10^6$/ mL) and/or motilities ($< 50\%$). The candidates were interviewed by a doctor, and those who smoked, had a tomato allergy, or had a history of relevant illness, such as adult-onset mumps orchitis, undescended testicles, or a high semen white blood cell (WBC) count ($\geq 1 \times 10^6$/mL) were excluded. Consequently, 54 subjects aged 26–50 (average: 36.9) years participated in the experiment.

7.6.2 STUDY DESIGN

Following a four-week observation period, the subjects were assigned by the Department of Pharmacy, IUHW Hospital into three groups: tomato juice group (n = 21), antioxidant group (n = 17), and control group (n = 16).

The subjects in the tomato juice group and the antioxidant group consumed one can of tomato juice or one capsule of an antioxidant pill daily, respectively, for 12 weeks (feeding period). The subjects in the control group were not administered any experimental foods and were required to avoid lycopene-rich foods containing tomatoes throughout the experimental period. Semen samples were collected at 0, 6, and 12 weeks, and blood samples were drawn at 0 and 12 weeks of the feeding period (Figure 7.4).

7.6.3 EXPERIMENTAL FOODS

A commercially available tomato juice containing 30 mg of lycopene, 38 mg of vitamin C, and 3 mg of vitamin E in one can was used as the experimental food for the tomato juice group. The capsules for the antioxidant group contained vitamin C (600 mg), vitamin E (200 mg), and glutathione (300 mg) (Figure 7.4).

7.6.4 MEASUREMENT

As the blood and semen antioxidant index, we measured the plasma and seminal plasma lycopene levels and seminal plasma MDA concentrations. As the semen parameter, we measured the semen volume, sperm concentration, sperm motility, abnormal sperm rate, seminal plasma WBC, straight-line velocity, curvilinear velocity, straightness, amplitude of the lateral head displacement, sperm head pitch, degree of sperm-nucleus damage, and degree of sperm-nucleus damage (Figure 7.4).

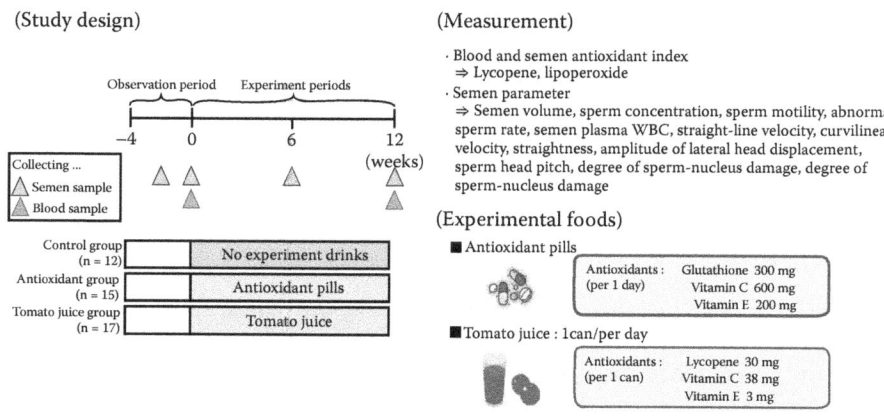

FIGURE 7.4 Study design, measurement, and experimental foods.

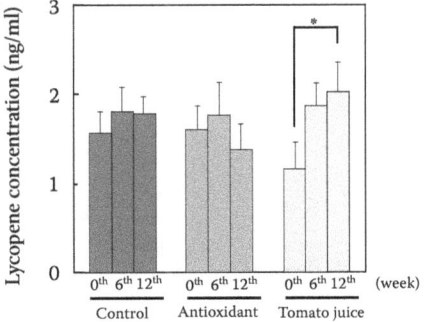
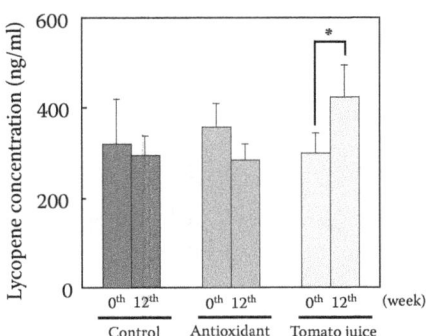

FIGURE 7.5   Mean lycopene levels in the plasma (left) and the seminal plasma (right) in the three groups. Data are presented as means ± SEMs. *p < 0.05 vs. 0th week

7.6.5   Results

7.6.5.1   Plasma and Seminal Plasma Lycopene Levels

Figure 7.5 shows the mean values and ranges of plasma and seminal plasma lycopene levels during the experimental period. The regular consumption of tomato juice for 12 weeks significantly increased the plasma lycopene level. The lycopene level in the seminal plasma was about 0.5% of that in the blood plasma. The seminal plasma lycopene levels also increased at the twelfth week in the tomato juice group.

7.6.5.2   Seminal Plasma MDA Concentration

The sample measurement was incomplete (control group: 4, antioxidant group: 8, tomato juice group: 6) because of a freezer problem. A decreasing trend with time was observed in the antioxidant and tomato juice groups but was not statistically significant.

7.6.5.3   Semen Parameter

There was no statistically significant difference in the parameters in the antioxidant or tomato juice groups compared with that in the control group. The changes in the sperm parameters from the beginning of the experimental period are summarized in Table 7.5. In the tomato juice group, the amount of decrease in the seminal plasma WBCs was statistically significant compared with that in the control group at the twelfth week ($P = 0.039$); an increase in the sperm motility was also statistically significant compared with that in the control group at the sixth week ($P = 0.019$). In the tomato juice and antioxidant groups, the amount of decrease in the semen volume was statistically very significant compared with that in the control group at the twelfth week (tomato juice group vs. control group; $P = 0.037$, antioxidant group vs. control group; $P = 0.035$).

7.6.5.4   Discussion

The consumption of tomato juice decreased the semen WBCs in the present study. The WHO defines leukocytospermia as semen WBCs $> 10^6$/mL. Although the

TABLE 7.5
Changes in the Semen Parameters in the Three Groups

Semen parameter	Control Group		Antioxidant Group		Tomato Juice Group	
	Week 6	Week 12	Week 6	Week 12	Week 6	Week 12
Semen volume (mL)	1.5 ± 2.5	0.2 ± 3.7	1.2 ± 1.3	−1.1 ± 0.7	−0.8 ± 1.5	0.3 ± 1.5
Sperm concentration (×10⁶/mL)	0.45 ± 0.28	0.40 ± 0.41	−0.01 ± 0.20	−0.48 ± 0.22*	−0.13 ± 0.23	−0.21 ± 0.20*
Sperm motility (%)	12.8 ± 20.6	−13.0 ± 9.46	5.38 ± 7.80	−11.9 ± 5.48	15.9 ± 7.29	0.64 ± 6.76
Abnormal sperm rate (%)	−3.24 ± 3.35	−0.87 ± 4.77	−2.28 ± 3.76	5.24 ± 5.28	5.87 ± 3.26*	0.08 ± 2.93
Seminal plasma WBC (×10⁶/mL)	0.89 ± 1.36	4.21 ± 2.13	−2.67 ± 1.74	−2.41 ± 1.46	0.82 ± 1.77	1.56 ± 1.46
Straight-line velocity (μm/s)	0.05 ± 0.15	−0.02 ± 0.16	0.02 ± 0.17	−0.03 ± 0.09	−0.01 ± 0.18	−0.21 ± 0.08*
Curvilinear velocity (μm/s)	−0.67 ± 2.51	−1.87 ± 1.00	0.63 ± 2.20	−1.02 ± 1.59	0.65 ± 1.36	1.86 ± 1.28
Straightness (%)	0.54 ± 4.72	−1.80 ± 2.9	2.86 ± 3.45	−2.84 ± 3.42	2.67 ± 2.74	3.27 ± 2.66
Amplitude of the lateral head displacement (Hz)	−5.33 ± 2.57	−2.16 ± 2.26	−0.91 ± 1.77	−1.66 ± 2.41	−1.27 ± 1.55	0.42 ± 1.87
Sperm head pitch (μm)	0.09 ± 0.12	−0.04 ± 0.06	0.09 ± 0.06	−0.06 ± 0.04	0.06 ± 0.05	0.42 ± 0.37
Degree of sperm-nucleus damage (%)	−0.96 ± 0.87	2.07 ± 0.86	0.19 ± 0.45	−0.87 ± 0.79*	−0.94 ± 0.52	−0.57 ± 0.55
Degree of sperm-nucleus damage (%)	10.19 ± 2.83	4.15 ± 2.50	2.30 ± 2.95	1.03 ± 2.58	2.72 ± 2.94	0.42 ± 2.94

Data are presented as means ± SEMs.
*$P < 0.05$ vs. control group.

association between semen WBCs and quality is still a matter of debate in the literature [116], many studies reported that the semen WBCs negatively affect the semen quality as a result of ROS produced by the WBCs [117–119]. In this study, the effects of the decrease in the semen WBCs on semen quality in the tomato juice group were uncertain because the baseline value was in the normal range ($0.63 \pm 0.08 \times 10^6$/mL).

We recruited subjects with low sperm concentrations ($< 20 \times 10^6$/mL) or low sperm motilities ($< 50\%$). After regular consumption of tomato juice, there was no statistically significant change in the sperm concentration; however, there was a significant improvement in the sperm motility at the sixth week. Various authors recognize semen parameters, such as sperm motility, concentration, and morphology, as vital for the assessment of fertility. Semen parameters were also correlated with *in vitro* oocyte fertilization rates [120–122]. Our results indicate that tomato juice consumption for six weeks may have a positive effect on asthenozoospermia by improving the sperm motility.

However, the improvement was not observed at the twelfth week in the same group. This might be dependent upon the large variation in the sperm motility within an individual, which is influenced by many factors, such as a period of sexual abstinence. The usefulness of tomato juice consumption for sperm motility should be evaluated in further prospective studies.

Gupta and Kumar reported that the consumption of 2 mg of lycopene twice a day improved the sperm motility in men with idiopathic nonobstructive oligo/astheno/teratozoospermia [108]. The amounts of vitamin C (38 mg) and vitamin E (3 mg) in the can of tomato juice that we used were lower than those in the antioxidant pills (vitamin E 200 mg/day, vitamin C 200 mg/day, glutathione 400 mg/day) in this study. This suggests that lycopene is the active ingredient in tomato juice that improves sperm motility.

7.6.5.5 Conclusion

In conclusion, regular consumption of tomato juice seems to improve the sperm motility in infertile patients. This is the first study to show that commercially available food, such as tomato juice, might benefit male infertility. The active ingredient in tomato juice that improves sperm motility may be lycopene; however, the mechanism is still unknown. To confirm the effect of tomato juice in detail, we are now planning a large-scale interventional study.

REFERENCES

1. Zegers-Hochschild F, Adamson GD, de Mouzon J, Ishihara O, Mansour R, Nygren K, Sullivan E, and Vanderpoel S; International Committee for Monitoring Assisted Reproductive Technology; World Health Organization. 2009. "International Committee for Monitoring Assisted Reproductive Technology (ICMART) and the World Health Organization (WHO) revised glossary of ART terminology, 2009." *Fertil Steril* 92(5):1520–4. Doi: 10.1016/j.fertnstert.2009.09.009.
2. USA: ORC Macro and the World Health Organization. 2004. "Infecundity, infertility, and childlessness in developing countries." DHS Comparative Reports No 9.
3. Mascarenhas MN, Flaxman SR, Boerma T, Vanderpoel S, and Stevens GA. 2012. "National, regional, and global trends in infertility prevalence since 1990: A systematic analysis of 277 health surveys." *PLoS Med* 9:e1001356.

4. Men's health—Male factor infertility. University of Utah Health Sciences Center. January 4, 2003. Archived from the Original on July 4, 2007.
5. Brugh VM 3rd and Lipshultz LI. 2004. "Male factor infertility: Evaluation and management." *Med Clin North Am* 88(2):367–85.
6. Lotti F and Maggi M. 2015. "Ultrasound of the male genital tract in relation to male reproductive health." *Hum Reprod Update* 21(1):56–83. Doi: 10.1093/humupd/dmu042.
7. World Health Organization. 1999. "WHO laboratory manual for the examination of human semen and semen-cervical mucus interaction." pp. 1–86.
8. Nieschlag E, Behre HM, and Van Ahlen H. 2000. *Andrology. Male reproductive health and dysfunction.* Springer, 2000.
9. Okada H, Japanese Male Hypogonadotropic Hypogonadism Study Group, O'Dea LS, Decosterd G, and Warrne D. 2006. "Combination therapy with recombinant human follicle-stimulating homone (follitropin alfa; GONALEF® and human chorionic gonadotropin is effective and well tolerated in azoospermic Japanese men with hypogonadotropic hypogonadism." *Hormone to Rinsyou* 54:725–32.
10. Esteves SC and Agarwal A. 2011. "Novel concepts in male infertility." *Int Braz J Urol* 37(1):5–15.
11. Agarwal A, Roychoudhury S, Bjugstad KB, and Cho CL. "Oxidation-reduction potential of semen: What is its role in the treatment of male infertility?" *Ther Adv Urol* 8(5):302–18.
12. Walczak-Jedrzejowska R, Wolski JK, and Slowikowska-Hilczer J. 2013. "The role of oxidative stress and antioxidants in male fertility." *Cent European J Urol* 66(1):60–7.
13. Pasqualotto FF, Sharma RK, Nelson DR, Thomas AJ, and Agarwal A. 2000. "Relationship between oxidative stress, semen characteristics and clinical diagnosis in men undergoing infertility investigation." *Fertil Steril* 73:459–64.
14. Iwasaki A and Gagnon C. 1992. "Formation of reactive oxygen species in spermatozoa of infertile patients." *Fertil Steril* 57:409.
15. Kao SH, Chao HT, Chen HW, Hwang TI, Liao TL, and Wei YH. 2008. "Increase of oxidative stress in human sperm with lower motility." *Fertil Steril* 89:1183–90.
16. de Lamirande E and Gagnon C. 1993. "A positive role for the superoxide anion in triggering hyperactivation and capacitation of human spermatozoa." *Int J Androl* 16:21–5.
17. Olugbenga OM, Olukole SG, Adeoye AT, and Adejoke AD. 2011. "Semen characteristics and sperm morphological studies of the West African Dwarf Buck treated with Aloe vera gel extract." *Iran J Reprod Med* 9:83–8.
18. Sanocka D and Kurpisz M. 2004. "Reactive oxygen species and sperm cells." *Reprod Biol Endocrinol* 23;2:12. Doi: 10.1186/1477-7827-2-12.
19. Rajes M, Damodar B, and Jitamanyu C. 2014. "Role of membrane lipid fatty acids in sperm cryopreservation." *Advances in Andrology* Volume. Article ID 190542, 9 pages.
20. Tavilani H, Doosti M, Abdi K, Vaisiraygani A, and Joshaghani HR. 2006. "Decreased polyunsaturated and increased saturated fatty acid concentration in spermatozoa from asthenozoospermic males as compared with normozoospermic males." *Andrologia* 38(5):173–8.
21. Hampl R, Drábková P, Kanďár R, and Stěpán J. 2012. "Impact of oxidative stress on male Infertility." *Ceska Gynekol* 77(3):241–5.
22. Zini A, Bielecki R, Phang D, and Zenzes MT. 2001. "Correlations between two markers of sperm DNA integrity, DNA denaturation and DNA fragmentation, in fertile and infertile men." *Fertil Steril* 75(4):674–7.
23. Kodama H, Yamaguchi R, Fukuda J, Kasai H, and Tanaka T. 1997. "Increased oxidative deoxyribonucleic acid damage in the spermatozoa of infertile male patients." *Fertil Steril* 68(3):519–24.
24. Spanò M, Bonde JP, Hjøllund HI, Kolstad HA, Cordelli E, and Leter G. 2000. "Sperm chromatin damage impairs human fertility. The Danish First Pregnancy Planner Study Team." *Fertil Steril* 73(1):43–50.

25. Irvine DS, Twigg JP, Gordon EL, Fulton N, Milne PA, and Aitken RJ. 2000. "DNA integrity in human spermatozoa: Relationships with semen quality." *J Androl* 21(1):33–44.
26. Muratori M, Piomboni P, Baldi E, Filimberti E, Pecchioli P, Moretti E, Gambera L, Baccetti B, Biagiotti R, Forti G, and Maggi M. 2000. "Functional and ultrastructural features of DNA-fragmented human sperm." *J Androl* 21(6):903–12.
27. Aitken RJ, De Iuliis GN, Finnie JM, Hedges A, and McLachlan RI. 2010. "Analysis of the relationships between oxidative stress, DNA damage and sperm vitality in a patient population: Development of diagnostic criteria." *Hum Reprod* 25(10):2415–26.
28. Barroso G, Morshedi M, and Oehninger S. 2000. "Analysis of DNA fragmentation, plasma membrane translocation of phosphatidylserine and oxidative stress in human spermatozoa." *Hum Reprod* 15(6):1338–44.
29. Schulte RT, Ohl DA, Sigman M, and Smith GD. 2010. "Sperm DNA damage in male infertility: Etiologies, assays, and outcomes." *J Assist Reprod Genet* 27(1):3–12.
30. Guz J, Gackowski D, Foksinski M, Rozalski R, Zarakowska E, Siomek A, Szpila A, Kotzbach M, Kotzbach R and Olinski R. 2013 "Comparison of oxidative stress/DNA damage in semen and blood of fertile and infertile men." *PLoS One* 8(7):e68490.
31. Aitken RJ and Baker MA. 2013. "Causes and consequences of apoptosis in spermatozoa; contributions to infertility and impacts on development." *Int J Dev Biol* 57(2–4):265–72.
32. Moustafa MH, Sharma RK, Thornton J, Mascha E, Abdel-Hafez MA, Thomas AJ Jr, and Agarwal A. 2004. "Relationship between ROS production, apoptosis and DNA denaturation in spermatozoa from patients examined for infertility." *Hum Reprod* 19(1):129–38.
33. Fathi Najafi T, Hejazi M, Feryal Esnaashari F, Sabaghiyan E, Hajibabakashani S, and Yadegari Z. 2012. "Assessment of sperm apoptosis and semen quality in infertile men-meta analysis." *Iran Red Crescent Med J* 14(3):182–3.
34. Doaa A and Eman ME. 2014. "Role of oxidative stress in male fertility and idiopathic infertility: Causes and treatment." *J Diagn Tech Biomed Anal* 3:1.
35. Goverde H, Dekker HS, Janssen H, Bastiaans BA, Rolland R, and Zielhuis GA. 1994. "Semen quality and frequency of smoking and alcohol consumption—An explorative study." *Int J Fertil Menopaus Stud* 40:135–8.
36. Agarwal A and Prabakaran SA. 2005. "Mechanism, measurement, and prevention of oxidative stress in male reproductive physiology." *Indian J Exp Biol* 43(11):963–74.
37. Sharma RK, Said T, and Agarwal A. 2004. "Sperm DNA damage and its clinical relevance in assessing reproductive outcome." *Asian J Androl* 6(2):139–48.
38. Alexander NJ. 1982. "Male evaluation and semen analysis." *Clin Obstet Gynecol* 25(3):463–82.
39. Talebi AR, Sarcheshmeh AA, Khalili MA, and Tabibnejad N. 2011. "Effects of ethanol consumption on chromatin condensation and DNA integrity of epididymal spermatozoa in rat." *Alcohol* 45(4):403–409.
40. Lavranos G, Balla M, Tzortzopoulou A, Syriou V, and Angelopoulou R. 2012. "Investigating ROS sources in male infertility: A common end for numerous pathways." *Reprod Toxicol* 34(3):298–307.
41. Mohammadi S, Jalali M, Nikravesh MR, Fazel A, Ebrahimzadeh A, Gholamin M, and Sankian M. 2013. "Effects of Vitamin-E treatment on CatSper genes expression and sperm quality in the testis of the aging mouse." *Iran J Reprod Med* 11(12):989–98.
42. Mostafa T, Tawadrous G, Roaia MM, Amer MK, Kader RA, and Aziz A. 2006. "Effect of smoking on seminal plasma ascorbic acid in infertile and fertile males." *Andrologia* 38(6):221–4.
43. Lee BM, Lee SK, and Kim HS. 1998. "Inhibition of oxidative DNA damage, 8-OHdG, and carbonyl contents in smokers treated with antioxidants (vitamin E, vitamin C, beta-carotene and red ginseng)." *Cancer Lett* 132:219–27.

44. Dietrich M, Block G, Norkus EP, Hudes M, Traber MG, Cross CE, and Packer L. 2003. "Smoking and exposure to environmental tobacco smoke decrease some plasma antioxidants and increase gamma-tocopherol in vivo after adjustment for dietary antioxidant intakes." *Am J Clin Nutr* 77(1):160–6.

45. Saleh RA, Agarwal A, Sharma RK, Nelson DR, and Thomas AJ, Jr. 2000. "Effect of cigarette smoking on levels of seminal oxidative stress in infertile men: A prospective study." *Fertil Steril* 78(3):491–9.

46. Künzle R, Mueller MD, Hänggi W, Birkhäuser MH, Drescher H, and Bersinger NA. 2003. "Semen quality of male smokers and nonsmokers in infertile couples." *Fertil Steril* 79(2):287–91.

47. Kiziler AR, Aydemir B, Onaran I, Alici B, Ozkara H, Gulyasar T, and Akyolcu MC. 2007. "High levels of cadmium and lead in seminal fluid and blood of smoking men are associated with high oxidative stress and damage in infertile subjects." *Biol Trace Elem Res* 120(1–3):82–91.

48. Vine MF, Tse CK, Hu P, and Truong KY. 1996. "Cigarette smoking and semen quality." *Fertil Steril* 65(4):835–42.

49. Jeng HA. 2014. "Exposure to endocrine disrupting chemicals and male reproductive health." *Front Public Health* 5;2:55.

50. Jurasović J, Cvitković P, Pizent A, Colak B, and Telisman S. 2004. "Semen quality and reproductive endocrine function with regard to blood cadmium in Croatian male subjects." *Biometals* 17(6):735–43.

51. Latini G, Del Vecchio A, Massaro M, Verrotti A, and De Felice C. 2006. "Phthalate exposure and male infertility." *Toxicology* 21;226(2–3):90–8.

52. Pant N, Shukla M, Kumar Patel D, Shukla Y, Mathur N, Kumar Gupta Y, and Saxena DK. 2008. "Correlation of phthalate exposures with semen quality." *Toxicol Appl Pharmacol* 15;231(1):112–16.

53. Lee E, Ahn MY, Kim HJ, Kim IY, Han SY, Kang TS, Hong JH, Park KL, Lee BM, and Kim HS. 2007. "Effect of di(n-butyl) phthalate on testicular oxidative damage and antioxidant enzymes in hyperthyroid rats." *Environ Toxicol* 22(3):245–55.

54. Feng M, Chen P, Wei X, Zhang Y, Zhang W, and Qi Y. 2011. "Effect of 4-nonylphenol on the sperm dynamic parameters, morphology and fertilization rate of Bufo raddei." *Afr J Biotechnol* 10:2698–707.

55. Ansoumane K, Peng D, Hady K, Aidogie O, Suqin Q, Chao Q, Tingting Y, and Kedi Y. 2015. "In vitro assessment of ROS on motility of epididymal sperm of male rat exposed to intraperitoneal administration of nonylphenol." *Asian Pac J Reprod* 4(3):169–78.

56. Agarwal A, Deepinder F, Sharma RK, Ranga G, and Li J. 2008. "Effect of cell phone usage on semen analysis in men attending infertility clinic: An observational study." *Fertil Steril* 89(1):124–8.

57. Aitken RJ, Bennetts LE, Sawyer D, Wiklendt AM, and King BV. "Impact of radio frequency electromagnetic radiation on DNA integrity in the male germline." *Int J Androl* 28(3):171–9.

58. De Iuliis GN, Newey RJ, King BV, and Aitken RJ. 2009. "Mobile phone radiation induces reactive oxygen species production and DNA damage in human spermatozoa in vitro." *PLoS One* 31;4(7):e6446.

59. Shiraishi K, Matsuyama H, and Takihara H. 2012. "Pathophysiology of varicocele in male infertility in the era of assisted reproductive technology." *Int J Urol* 19(6):538–50.

60. French DB, Desai NR, and Agarwal A. 2008. "Varicocele repair: Does it still have a role in infertility treatment?" *Curr Opin Obstet Gynecol* 20(3):269–74.

61. Fischer MA, Willis J, and Zini A. 2003. "Human sperm DNA integrity: Correlation with sperm cytoplasmic droplets." *Urology* 61(1):207–11.

62. Barbieri ER, Hidalgo ME, Venegas A, Smith R, and Lissi EA. 1999. "Varicocele-associated decrease in antioxidant defenses." *J Androl* 20(6):713–17.

63. Meucci E, Milardi D, Mordente A, Martorana GE, Giacchi E, De Marinis L, and Mancini A. 2003. "Total antioxidant capacity in patients with varicoceles." *Fertil Steril* 79 Suppl 3:1577–83.
64. Chehval MJ and Purcell MH. 1999. "Deterioration of semen parameters over time in men with untreated varicocele: Evidence of progressive testicular damage." *Fertil Steril* 57(1):174–7.
65. Kasturi SS, Tannir J, and Brannigan RE. 2008. "The metabolic syndrome and male infertility." *J Androl* 29(3):251–9.
66. Bakos HW, Mitchell M, Setchell BP, and Lane M. 2011. "The effect of paternal diet-induced obesity on sperm function and fertilization in a mouse model." *Int J Androl* 34(5 Pt 1):402–10.
67. Kashou AH, du Plessis SS, and Agarwal A. 2012. *The role of obesity in ROS generation and male infertility. Studies on Men's Health and Fertility.* Springer, 2012. pp. 571–90.
68. Furukawa S, Fujita T, Shimabukuro M, Iwaki M, Yamada Y, Nakajima Y, Nakayama O, Makishima M, Matsuda M, and Shimomura I. 2004. "Increased oxidative stress in obesity and its impact on metabolic syndrome." *J Clin Invest* 114(12):1752–61.
69. Mallidis C, Agbaje I, McClure N, and Kliesch S. 2011. "The influence of diabetes mellitus on male reproductive function: A poorly investigated aspect of male infertility." *Urologe A* 50(1):33–7.
70. Bloch-Damti A and Bashan N. 2005. "Proposed mechanisms for the induction of insulin resistance by oxidative stress." *Antioxid Redox Signal* 7(11–12):1553–67.
71. Amaral S, Oliveira PJ, and Ramalho-Santos J. 2008. "Diabetes and the impairment of reproductive function: Possible role of mitochondria and reactive oxygen species." *Curr Diabetes Rev* 4(1):46–54.
72. Shrilatha B and Muralidhara. 2007. "Early oxidative stress in testis and epididymal sperm in streptozotocin-induced diabetic mice: Its progression and genotoxic consequences." *Reprod Toxicol* 23(4):578–87.
73. Shrilatha B and Muralidhara. 2007. "Occurrence of oxidative impairments, response of antioxidant defences and associated biochemical perturbations in male reproductive milieu in the streptozotocin-diabetic rat." *Int J Androl* 30(6):508–18.
74. Amaral S, Oliveira PJ, and Ramalho-Santos J. 2008. "Diabetes and the impairment of reproductive function: Possible role of mitochondria and reactive oxygen species." *Curr Diabetes Rev* 4(1):46–54.
75. Brownlee M. 2005. "The pathobiology of diabetic complications: A unifying mechanism." *Diabetes* 54(6):1615–25.
76. Agbaje IM, Rogers DA, McVicar CM, McClure N, Atkinson AB, Mallidis C, and Lewis SE. 2007. "Insulin-dependent diabetes mellitus: Implications for male reproductive function." *Hum Reprod* 22(7):1871–7.
77. Sabeti P, Pourmasumi S, Rahiminia T, Akyash F, and Talebi AR. 2016. "Etiologies of sperm oxidative stress." *Int J Reprod Biomed (Yazd)* 14(4):231–40.
78. Suleiman SA, Ali ME, Zaki ZM, el-Malik EM, and Nasr MA. 1996. "Lipid peroxidation and human sperm motility: Protective role of vitamin E." *J Androl* 17(5):530–7.
79. Kessopoulou E, Powers HJ, Sharma KK, Pearson MJ, Russell JM, Cooke ID, and Barratt CL. 1995. "A double-blind randomized placebo cross-over controlled trial using the antioxidant vitamin E to treat reactive oxygen species associated male infertility." *Fertil Steril* 64(4):825–31.
80. Showell MG, Brown J, Yazdani A, Stankiewicz MT, and Hart RJ. 2011. "Antioxidants for male subfertility." *Cochrane Database Syst Rev* 19;(1):CD007411.
81. Dawson EB, Harris WA, Teter MC, and Powell LC. 1992. "Effect of ascorbic acid supplementation on the sperm quality of smokers." *Fertil Steril* 58(5):1034–9.

82. Kessopoulou E, Powers HJ, Sharma KK, Pearson MJ, Russell JM, Cooke ID, and Barratt CL. 1995. "A double-blind randomized placebo cross-over controlled trial using the antioxidant vitamin E to treat reactive oxygen species associated male infertility." *Fertil Steril* 64(4):825–31.

83. Moilanen J and Hovatta O. 1995. "Excretion of alpha-tocopherol into human seminal plasma after oral administration." *Andrologia* 27(3):133–6.

84. Rolf C, Cooper TG, Yeung CH, and Nieschlag E. 1999. "Antioxidant treatment of patients with asthenozoospermia or moderate oligoasthenozoospermia with high-dose vitamin C and vitamin E: A randomized, placebo-controlled, double-blind study." *Hum Reprod* 14(4):1028–33.

85. Gülçin I. 2006. "Antioxidant and antiradical activities of L-carnitine." *Life Sci* 78(8):803–11.

86. Sener G, Paskaloğlu K, Satiroglu H, Alican I, Kaçmaz A, and Sakarcan A. 2004. "L-carnitine ameliorates oxidative damage due to chronic renal failure in rats." *J Cardiovasc Pharmacol* 43(5):698–705.

87. Balercia G, Regoli F, Armeni T, Koverech A, Mantero F, and Boscaro M. 2005. "Placebo-controlled double-blind randomized trial on the use of L-carnitine, L-acetylcarnitine, or combined L-carnitine and L-acetylcarnitine in men with idiopathic asthenozoospermia." *Fertil Steril* 84(3):662–71.

88. Lenzi A, Sgrò P, Salacone P, Paoli D, Gilio B, Lombardo F, Santulli M, Agarwal A, and Gandini L. 2004. "A placebo-controlled double-blind randomized trial of the use of combined l-carnitine and l-acetyl-carnitine treatment in men with asthenozoospermia." *Fertil Steril* 81(6):1578–84.

89. Peivandi S, Abasali K, and Narges M. 2010 "Effects of L-carnitine on infertile men's spermogram: A randomised clinical trial." *J Reprod Infertil* 10:331.

90. Zhou X and Liu F, Zhai S. 2007. "Effect of L-carnitine and/or L-acetyl-carnitine in nutrition treatment for male infertility: A systematic review." *Asia Pac J Clin Nutr* 16 Suppl 1:383–90.

91. Lenzi A, Sgrò P, Salacone P, Paoli D, Gilio B, Lombardo F, Santulli M, Agarwal A, and Gandini L. 2004. "A placebo-controlled double-blind randomized trial of the use of combined l-carnitine and l-acetyl-carnitine treatment in men with asthenozoospermia." *Fertil Steril* 81(6):1578–84.

92. Trevisan R, Flesch S, Mattos JJ, Milani MR, Bainy AC, and Dafre AL. 2014. "Zinc causes acute impairment of glutathione metabolism followed by coordinated antioxidant defenses amplification in gills of brown mussels Perna perna." *Comp Biochem Physiol C Toxicol Pharmacol* 159:22–30.

93. Gavella M, Lipovac V, Vucić M, and Sverko V. 1999. "In vitro inhibition of superoxide anion production and superoxide dismutase activity by zinc in human spermatozoa." *Int J Androl* 22(4):266–74.

94. Türk S, Mändar R, Mahlapuu R, Viitak A, Punab M, and Kullisaar T. 2014. "Male infertility: Decreased levels of selenium, zinc and antioxidants." *J Trace Elem Med Biol* 28(2):179–85.

95. Omu AE, Dashti H, and Al-Othman S. 1998. "Treatment of asthenozoospermia with zinc sulphate: Andrological, immunological and obstetric outcome." *Eur J Obstet Gynecol Reprod Biol* 79(2):179–84.

96. Wong WY, Merkus HM, Thomas CM, Menkveld R, Zielhuis GA, and Steegers-Theunissen RP. 2002. "Effects of folic acid and zinc sulfate on male factor subfertility: A double-blind, randomized, placebo-controlled trial." *Fertil Steril* 77(3):491–8.

97. Lombardo F, Sansone A, Romanelli F, Paoli D, Gandini L, and Lenzi A. 2011. "The role of antioxidant therapy in the treatment of male infertility: An overview." *Asian J Androl* 13(5):690–7.

98. Di Mascio P, Kaiser S, and Sies H. 1989. "Lycopene as the most efficient biological carotenoid singlet oxygen quencher." *Arch Biochem Biophys* 274(2):532–8.

99. Kong KW, Khoo HE, Prasad KN, Ismail A, Tan CP, and Rajab NF. 2010. "Revealing the power of the natural red pigment lycopene." *Molecules* 15(2):959–87.
100. Rao AV and Rao LG. 2007. "Carotenoids and human health." *Pharmacol Res* 55(3):207–16.
101. Stahl W, Schwarz W, Sundquist AR, and Sies H. 1992. "Cis-trans isomers of lycopene and beta-carotene in human serum and tissues." *Arch Biochem Biophys* 294(1):173–7.
102. Palan P and Naz R. 1996. "Changes in various antioxidant levels in human seminal plasma related to immunoinfertility." *Arch Androl* 36(2):139–43.
103. Goyal A, Chopra M, Lwaleed BA, Birch B, and Cooper AJ. 2007. "The effects of dietary lycopene supplementation on human seminal plasma." *BJU Int* 99(6):1456–60.
104. Hekimoglu A, Kurcer Z, Aral F, Baba F, Sahna E, and Atessahin A. 2009. "Lycopene, an antioxidant carotenoid, attenuates testicular injury caused by ischemia/reperfusion in rats." *Tohoku J Exp Med* 218(2):141–7.
105. Mangiagalli MG, Cesari V, Cerolini S, Luzi F, and Toschi L. 2012. "Effect of lycopene supplementation on semen quality and reproductive performance in rabbit." *World Rabbit Sci* 20:141–8.
106. Mangiagalli MG, Martino PA, Smajlovic T, Guidobono Cavalchini L, and Marelli SP. "Effect of lycopene on semen quality, fertility and native immunity of broiler breeder." *Br Poult Sci* 51(1):152–7.
107. Aly HA, El-Beshbishy HA, and Banjar ZM. 2012. "Mitochondrial dysfunction induced impairment of spermatogenesis in LPS-treated rats: Modulatory role of lycopene." *Eur J Pharmacol* 677(1–3):31–8.
108. Gupta NP and Kumar R. 2002. "Lycopene therapy in idiopathic male infertility—A preliminary report." *Int Urol Nephrol* 34(3):369–72.
109. Mohanty N, Kumar S, Jha A, and Arora R. 2001. "Management of idiopathic oligoasthenospermia with lycopene." *Indian J Urol* 18:57–61.
110. Sarkar PD, Gupt T, and Sahu A. "Comparative analysis of lycopene in oxidative stress." *J Assoc Physicians India* 60:17–19.
111. Zini A, San Gabriel M, and Libman J. 2010. "Lycopene supplementation in vitro can protect human sperm deoxyribonucleic acid from oxidative damage." *Fertil Steril* 94(3):1033–6.
112. Rosato MP, Centoducati G, Santacroce MP, and Iaffaldano N. 2012. "Effects of lycopene on in vitro quality and lipid peroxidation in refrigerated and cryopreserved turkey spermatozoa." *Br Poult Sci* 53(4):545–52.
113. Breinholt V, Lauridsen ST, Daneshvar B, and Jakobsen J. 2000. "Dose-response effects of lycopene on selected drug-metabolizing and antioxidant enzymes in the rat." *Cancer Lett* 154(2):201–10.
114. Velmurugan B, Bhuvaneswari V, Burra UK, and Nagini S. 2002. "Prevention of N-methyl-N′-nitro-N-nitrosoguanidine and saturated sodium chloride-induced gastric carcinogenesis in Wistar rats by lycopene." *Eur J Cancer Prev* 11(1):19–26.
115. Ben-Dor A, Steiner M, Gheber L, Danilenko M, Dubi N, Linnewiel K, Zick A, Sharoni Y, and Levy J. 2005. "Carotenoids activate the antioxidant response element transcription system." *Mol Cancer Ther* 4(1):177–86.
116. Lackner JE. 2010. "The association between leukocytes and sperm quality is concentration dependent." *Reprod Biol Endocrinol* 8:12.
117. Aitken RJ, West K, and Buckingham D. 1994. "Leukocytic infiltration into the human ejaculate and its association with semen quality, oxidative stress, and sperm function." *J Androl* 15:343–52.
118. Whittington K, Harrison SC, Williams KM, Day JL, and McLaughlin EA. 1999. "Reactive oxygen species (ROS) production and the outcome of diagnostic tests of sperm function." *Int J Androl* 22:236–42.

119. Sharma RK, Pasqualotto AE, Nelson DR, Thomas AJ Jr, and Agarwal A. 2001. "Relationship between seminal white blood cell counts and oxidative stress in men treated at an infertility clinic." *J Androl* 22:575–83.
120. Oehninger S and Kruger T. 1995. "The diagnosis of male infertility by semen quality. Clinical significance of sperm morphology assessment." *Hum Reprod* 10:1037–8.
121. Enginsu ME, Dumoulin JC, Pieters MH, Evers JL, and Geraedts JP. "Predictive value of morphologically normal sperm concentration in the medium for in-vitro fertilization." *Int J Androl* 16:113–20.
122. Mashiach R, Fisch B, Eltes F, Tadir Y, Ovadia J, and Bartoor B. 1992. "The relationship between sperm ultrastructural features and fertilizing capacity in vitro." *Fertil Steril* 57:1052–7.

8 Lycopene and Tomatoes in the Prevention and Management of Other Human Diseases

A. Venketeshwer Rao and Leticia G. Rao

CONTENTS

8.1 INTRODUCTION

Scientific curiosity surrounding lycopene goes back several decades, when organic chemists established the molecular structure of lycopene and its antioxidant properties. Recognition of its biological significance followed a few years later and continues to interest scientists from all different disciplines. In 1995, a study from Harvard School of Medicine [1] suggested that lycopene intake may be related to reduced incidence of prostate cancer. This finding initiated a number of studies looking into the role of lycopene as a potent natural antioxidant in the prevention and management of several human diseases. These studies were based on the hypothesis that oxidative damage is caused by several environmental and lifestyle factors, including diet, leading to the accumulation of reactive oxygen species (ROS) and other free radicals responsible for cellular damage and the causation and progression of human diseases. Although the initial focus of the research was directed at the role of lycopene in the prevention of prostate cancer, several studies soon followed into its role in the prevention of other cancers, in particular breast and ovarian cancer, followed

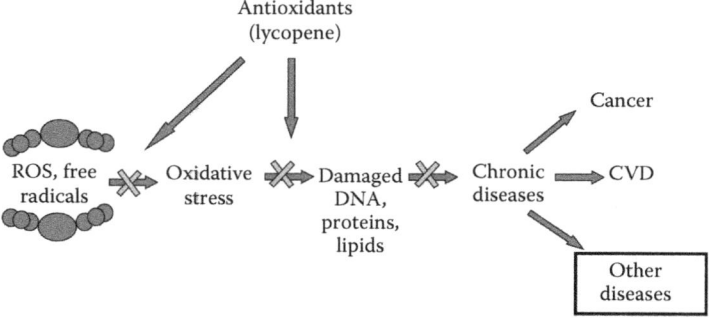

FIGURE 8.1 Role of oxidative stress in human diseases other than cancer and cardiovascular disease and where antioxidants such as lycopene play a role in mitigating the damaging effect of oxidative stress

---

TABLE 8.1
Role of Lycopene in Other Human Diseases

Respiratory diseases
Hypertension
Male infertility
Ultraviolet induced sunburns and other skin diseases
Neurodegenerative and Mental disorders
Inflammatory disorders – Rheumatoid arthritis
Bone diseases – Osteoporosis
Ocular diseases
Oral disorders

---

by cardiovascular disease (CVD). Several research articles and reviews have since been published addressing these specific human diseases.

More recently, based on the hypothesis that antioxidants such as lycopene can mitigate the damaging effects of oxidative stress on cellular DNA, proteins, and lipids (Figure 8.1), researchers have begun to study the role of lycopene in diseases other than cancer and CVD that are covered adequately in other chapters of this book. The focus of this chapter is to review recent studies investigating the role of lycopene in human diseases other than cancer and CVD.

Recent studies have documented the beneficial role of lycopene in the following human health diseases other than cancer and CVD (Table 8.1). However, lycopene is not limited only to the health disorders listed in Table 8.1. Future studies will undoubtedly evaluate its role in other important human diseases as well.

8.2 CHRONIC RESPIRATORY DISEASES (CRD)

According to the Public Health Agency of Canada, chronic respiratory diseases are chronic diseases of the airways and other parts of the lung. Some of the most common are asthma, chronic obstructive pulmonary disease (COPD), lung cancer, cystic fibrosis, sleep apnea, and occupational lung diseases [2]. Of all the diseases associated with the respiratory system, asthma has been the most studied. It is a lung disease caused by inflammation that leads to wheezing, shortness of breath, chest tightness, and coughing [2]. Asthma affects a substantial proportion of children and adults worldwide. Although the cause of asthma is not fully understood, it is believed that diets lacking in antioxidants leading to the accumulation of ROS and oxidative stress might play an important role in the modulation of respiratory diseases [3]. There is evidence of an oxidant-antioxidant imbalance in asthma [4]. Antioxidant-rich diets are therefore associated with reduced asthma prevalence in epidemiologic studies. Antioxidant rich foods such as fruit and vegetables are good sources of several carotenoids, including lycopene. Evidence suggests that reduced intake of fruit and vegetables may play a critical role in the development of asthma and allergies [5]. Studies, therefore, have been directed towards understanding the role of antioxidants and in particular carotenoids such as lycopene in the prevention and management of asthma.

A recent study evaluated the association between a high-antioxidant diet and the reduced prevalence of asthma. A total of 137 asthmatic adults were recruited into the study. Forty-six subjects were placed on a high-antioxidant diet and 91 subjects on a low-antioxidant diet for a period of 14 days. Following this, a parallel, randomized, control supplementation trial was initiated. Subjects on high-antioxidant diet served as the placebo group, while those on a low-antioxidant diet received either no supplementation or 45 mg lycopene per day for 14 weeks. Subjects consuming the low-antioxidant diet had lower forced expiratory volume and forced vital capacity than the group consuming the high-antioxidant diet. Also, the time to exacerbation was greater in the low-antioxidant diet compared to the high-antioxidant diet group, and they were 2.26 times as likely to exacerbate. Supplementation of the low-antioxidant diet with lycopene resulted in improvement in lung function [6].

In a study by Wood et al. [7], 30 asthmatic adult subjects were placed on a low-lycopene placebo and lycopene-supplemented diets (45 mg/d) for seven days. With the consumption of a low-lycopene diet the plasma carotenoid levels and asthma control scores both decreased. On the other hand, treatment with lycopene reduced airway neutrophil influx. Based on these results, the authors concluded that clinical asthma outcomes may be modified by dietary antioxidant consumption. They further suggested that changing dietary antioxidant intakes may be contributing to rising asthma prevalence.

In another study [8] peripheral blood samples were collected from 15 asthmatic and 16 healthy subjects and their dietary carotenoid levels, including lycopene, were estimated. Despite similar dietary intakes the whole blood levels of lycopene and other carotenoids were significantly lower in the asthmatic subjects compared to the control subjects. In a related study, nine healthy control subjects were supplemented with 20 mg per day of lycopene and other carotenoids. Although no significant increases were found in the serum and sputum samples after lycopene supplementation, changes in airway lycopene levels correlated significantly with changes in plasma levels. It was suggested that increasing plasma lycopene levels will also increase airway lycopene levels and provide protection against oxidative stress and the incidence of asthma.

In a recent study from Japan [9], correlation between various antioxidant nutrients in the plasma of Japanese patients and the incidence of chronic obstructive pulmonary disease (COPD), asthma-COPD overlap syndrome (ACOS), and bronchial asthma (BA) were compared to a healthy elderly control group. Significantly lower levels of plasma lycopene and other antioxidants were observed in the COPD patients compared to controls but not in the ACOS and BA patients. These results suggest that the development of COPD may be partly related to the failure of antioxidant nutrients such as lycopene and ascorbic acid to protect lung tissue from oxidative damage. A strategy to increase the levels of these antioxidants may be helpful in preventing the incidence of COPD.

Using an animal model, researchers exposed SAMR1 mice to cigarette smoke long enough to induce emphysema. The mice were then fed lycopene given as tomato juice mixed with water. They found that feeding tomato juice containing lycopene completely prevented smoke-induced emphysema. This effect was associated with the strong antioxidant properties of lycopene [10].

In summary, there is compelling evidence to suggest that accumulation of reactive oxygen species and oxidative stress plays an important role in the etiology of asthma and other respiratory diseases. Diets rich in sources of lycopene such as tomatoes and tomato products can play an important role in preventing respiratory diseases. Further human clinical studies in the future will help confirm the beneficial role of tomatoes, tomato products, and lycopene, as well as their mechanism of action.

8.3 HYPERTENSION

Blood pressure is the pressure of the blood in the circulatory system. If the pressure is too high, it puts extra strain on the arteries, leading to heart attacks and strokes. It is measured in "millimeters of mercury" (mmHg) and reported as systolic pressure over diastolic pressure. According to World Health Federation, high blood pressure is defined as a systolic blood pressure above 140 mmHg and/or a diastolic blood pressure above 90 mmHg. High blood pressure, also known as hypertension, is a condition commonly associated with narrowing of the arteries. It is the single most important risk factor for stroke. Although the exact cause of hypertension is unknown, there are several factors and conditions that may contribute to its occurrence, including genetic factors, family history of hypertension, obesity, sedentary lifestyle, excess salt intake, alcohol, smoking, stress, age, hormone levels, and abnormalities in the nervous and circulatory systems and kidneys [11–14].

Hypertensive subjects have been observed to have higher levels of reactive ROS and to be subjected to oxidative stress, leading scientists to suggest a causal relationship between oxidative stress and hypertension [11]. It was therefore postulated that antioxidants may play an important role in the prevention of hypertension. Since the recognition of lycopene as a potent antioxidant and its preventive role in oxidative stress-mediated chronic diseases, researchers are beginning to investigate its role in the prevention and management of hypertension.

A dietary approach to stop hypertension (DASH) diet is recommended for lowering high blood pressure [15]. The DASH diet was designed to give beneficial levels of fiber and other nutrients. As such, it contains more fruit, vegetables, and whole grains compared to control diets and is substantially higher in antioxidant phytochemicals. When the DASH diet was compared with the control diet, it was found to contain substantially higher levels of lycopene and other carotenoids, polyphenols, flavanols, flavanones, and flavan-3-ols. The beneficial effects of these phytochemicals in the management of blood pressure are now being recognized.

Hypertension is also associated with liver cirrhosis. In a study by Rocchi et al. [16], patients with liver cirrhosis were compared to healthy matched controls. A significant reduction in serum lycopene, other carotenoid antioxidants, retinol, and α-tocopherol were observed in the cirrhotic patients. In view of the link between hypertension and liver cirrhosis, a thorough screening for the antioxidants and improved diet in the follow-up of liver cirrhosis patients was recommended.

In a previously reported study, Paran and Engelhard [17] studied the effect of tomato lycopene on blood pressure using a single-blind, placebo-controlled trial. Thirty grade-one hypertensive patients between the ages of 40 and 65 years, not requiring any blood pressure and lipid-lowering medications, were recruited into the study. After a two-week run-in period for baseline evaluation, the patients were placed on four-week placebo and eight-week treatment periods. The treatment consisted of ingesting tomato extract dietary supplement capsules that provided 15 mg lycopene every day. Although the results showed no significant changes in the diastolic blood pressure after lycopene treatment, they did show a considerable reduction in the systolic blood pressure from the baseline value of 144 mm Hg to 134 mm Hg at the end of the lycopene treatment. Observations from other reported studies also support these observations [18–20]. (See Figure 8.2.)

In a similar study, Paran et al. [21] evaluated the effect of adding tomato extract on plasma lycopene concentration and nitrite and nitrate levels in 50 moderate hypertensive subjects being treated with one or two antihypertensive drugs. A significant decrease in systolic and diastolic blood pressures were observed after a six-week treatment with tomato extract supplementation. No differences in the blood pressure measurements were observed during the placebo period. However, serum lycopene levels were observed to increase with the tomato extract treatment and a significant inverse relationship was observed between systolic pressure and serum lycopene level.

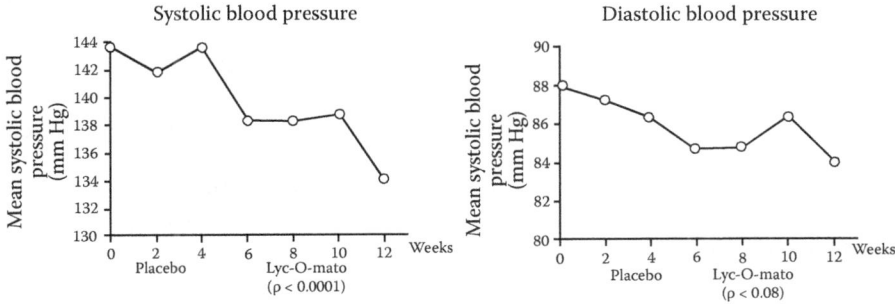

FIGURE 8.2   Effect of tomato extract on systolic and diastolic blood pressure. (Reprinted from *Adv Food Nutri Res*, 51, Rao A. V., L. G. Rao, Ray, M. R. Lycopene, 99–164, Copyright 2006, with permission from Elsevier.)

In another study [22], a significant reduction in plasma lycopene was observed in subjects with mild essential hypertension compared to normal subjects. Similar reductions in ascorbate, urate, and β-carotene were also observed in this study. However, there were no differences in the nitrous oxide derivatives between the two groups.

In a recent study [23] a control group of 50 healthy subjects who were non-smokers and having no chronic systemic illness were compared with a group of 40 grade-one hypertensive patients, to study the beneficial effect of lycopene from cooked tomatoes for 60 days. Lipid peroxidation rate, activity of plasma enzymes involved in antioxidant activities, and serum lipid profile were measured. Significantly lower plasma antioxidant enzyme activity, very high lipid peroxidation rate, and very high serum total cholesterol and triglycerides were observed in the grade-one hypertensive group when compared to the control group. Sixty days of tomato supplementation in the hypertensive group showed a significant improvement in the levels of serum enzymes involved in antioxidant activities and decreased lipid peroxidation rate.

A recent meta-analysis of lycopene intervention studies on blood pressure showed that a higher dosage of lycopene supplement (> 12 mg/d) could lower systolic blood pressure more significantly, especially for participants with a baseline systolic blood pressure of > 120 mmHg. Lycopene intervention had no statistical effect on diastolic blood pressure. The authors concluded that lycopene supplementation of 12 mg per day may effectively decrease systolic blood pressure [24]. However, in another epidemiological study, circulating carotenoid concentrations and the incidence of hypertension was investigated [25]. In this study the association of circulating carotenoids, a sum of four serum carotenoids (alpha-carotene, beta-carotene, lutein/zeaxanthin, and cryptoxanthin) and of lycopene with hypertension was investigated over a period of 20 years. Lycopene was observed to be unrelated to hypertension. However, individuals with higher concentrations of the sum of carotenoids, not including lycopene, generally had a lower risk for future hypertension.

Another risk factor for hypertension and therefore for cardiovascular disease is endothelial dysfunction due to oxidative stress [26] leading to impaired bioavailability and bioactivity of nitrous oxide (NO) that plays an important role in regulating hypertension. Animal studies have shown that disruption of endothelial NO synthase can lead to elevated levels of blood pressure compared to a control; animals. Clinical studies have also shown that inhibition of NO raises blood pressure. Arterial stiffness resulting in systolic hypertension has also been shown to be associated with impaired NO activity [27]. In a recent study [28] using cultured human umbilical endothelial cells, the effects of lycopene on the synthesis and release of NO and endothelial cell migration were studied. Results showed that lycopene increases NO generation and its release. Authors conclude that lycopene may be important in the management of NO-related diseases such as hypertension and angiogenesis.

In summary, epidemiological studies have shown that an intake of 12 mg per day of lycopene can have a protective effect against systolic blood pressure. Human intervention studies by and large have also supported this evidence. However, it should be pointed out that only a few controlled human clinical trials have been reported in the literature. More research is needed not only to provide strong support for the role of lycopene in the prevention of hypertension but also the mechanisms of action.

8.4   MALE INFERTILITY

Infertility is an important male reproductive disorder globally, leading to failure to achieve pregnancy in female partners. It is estimated that about 7–10% of adult men in their reproductive years are infertile and 25% of all men with infertility will have nonspecific or idiopathic infertility [29–31]. There are several factors that can contribute to male infertility. One of the main components of spermatozoa is the presence of polyunsaturated fatty acids in the lipid membrane, which makes them more prone to oxidative damage by ROS, resulting in the disruption of the fluidity of the spermatozoa cell membrane and reducing sperm motility and viability. Another effect of ROS on spermatozoa is DNA damage leading to complications of fertilization, implantation, and embryonic development [32].

The low sperm count observed in idiopathic male factor infertility may also be due to apoptosis of the spermatozoa [33]. Significant levels of ROS are detectable in the semen of up to 25% of infertile men, whereas fertile men do not produce detectable levels of ROS in their semen [14,34]. This observation led scientists to the suggestion that excessive amounts of ROS can cause structural damage to the sperm deoxyribonucleic acid, reduce motility, and damage the sperm membrane, leading to abnormal semen parameters and perhaps infertility [35]. This led to the possibility that preventing the damaging effect of ROS by antioxidants, sperm damage can be prevented and its functionality maintained (Figure 8.3).

Animal and human intervention studies were initiated to evaluate antioxidant vitamins such as vitamins C and E, mostly as single agents, on sperm quality and functionality. In general, these early studies showed improved sperm quality with antioxidant vitamin supplementation. Glutathione and l-carnitin are other important antioxidants that were also shown to improve sperm quality [36].

Since the recognition of lycopene as a potent antioxidant, and its preventive role in oxidative stress-mediated chronic diseases, studies were conducted to investigate its

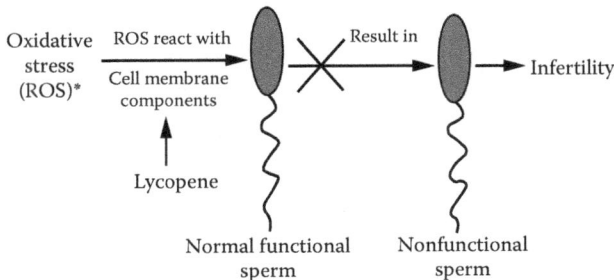

*Evidence to suggest that infertile men have high levels of ROS

FIGURE 8.3   Effect of oxidative stress and lycopene on sperm functionality. (Reprinted from *Adv Food Nutri Res*, 51, Rao A. V., L. G. Rao, Ray, M. R. Lycopene, 99–164, Copyright 2006, with permission from Elsevier.)

role in protecting sperm from oxidative damage and infertility in men. Palan and Naz [37] were the early researchers to show that men with antibody-mediated infertility had lower serum lycopene levels compared to fertile control subjects. In another study [38], 50 infertile male volunteers between the ages of 21 and 50 years were recruited. The subjects had a normal hormonal profile of antisperm antibody titre and no history of having taken any therapy for infertility or having obstructive azospermia. They consumed a daily dose of 8 mg lycopene in capsule form. The treatment was continued until sperm analysis showed optimal levels or until pregnancy of their partners was achieved. After a 12-month follow-up, it was reported that lycopene treatment resulted in significant increases in serum lycopene concentration. Significant improvements were also observed in other sperm quality indicators. The partners of 18 of the 50 subjects had successful pregnancies, accounting for a 36% success rate (Figure 8.4).

Durairajanayagam et al. [39] recently published a review article on lycopene and male fertility. In all they reviewed four studies relating to sperm count, sperm motility in five studies, sperm viability in only one study, and three studies relating to sperm morphology. Although there was a degree of contradiction in the reported results, in general lycopene was reported to increase sperm count, motility, and viability. Sperm morphology was also shown to improve with lycopene supplementation. Based on these observations the authors conclude that lycopene supplementation of 4–8 mg per day for 3–12 months was sufficient to treat male infertility. They further state that more research and clinical trials have to be conducted on humans to determine the most accurate therapeutic dosage.

In another study, several biologically active substances with antioxidant properties, such as vitamins C and E, quercetin, tannic acid, resveratrol, curcumin, and lycopene were investigated for their protective effect on sperm functionality. Lycopene was found to be the most highly efficient antioxidant and free radical scavenger, having a protective effect on spermatozoa [40].

Male infertility has also been associated with an imbalance in the ratio of the polyunsaturated fatty acids arachidonic acid and docosahexaenoic acid (AA/DHA). Filipcikova et al. [41] conducted a study to investigate the use of lycopene in modulating AA/DHA ratio and its correlation with male infertility. Three months of

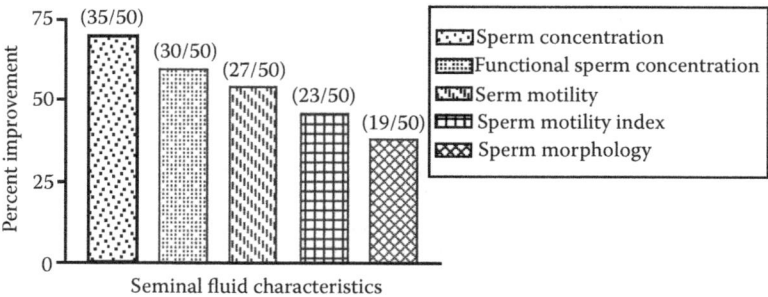

FIGURE 8.4 Effect of lycopene on the sperm quality in infertile men. (Reprinted from *Adv Food Nutri Res*, 51, Rao A. V., L. G. Rao, Ray, M. R. Lycopene, 99–164, Copyright 2006, with permission from Elsevier.)

treatment with lycopene led to a significant improvement in the AA/DHA ratio in seminal plasma of males from infertile couples and facilitated spontaneous as well as *in vitro fertilization* (IVF) conception.

In a more recent study, Yu et al. [42] investigated the effect of tomato juice consumption on seminal plasma lycopene levels and sperm parameters in infertile men with poor sperm concentration and motility. Subjects were assigned to a tomato juice group, an antioxidant group, and a control group. The tomato juice group consumed one can of tomato juice, containing 30 mg of lycopene, every day while the antioxidant group consumed one antioxidant capsule containing 600 mg of vitamin C, 200 mg of vitamin E, and 300 mg of glutathione every day for a period of 12 weeks. The control group consumed neither lycopene nor other antioxidant supplements. In the tomato juice group, there was a significant increase in plasma lycopene and sperm motility compared to the control group. No improvement was observed in the antioxidant group. The authors concluded that consuming a commercially available food such as tomato juice can improve sperm functionality in infertile men.

In summary, since the recognition of the role of oxidative stress in male infertility and the possible role that antioxidants can play in the treatment of male infertility, several animal and human intervention studies have been undertaken. With the recognition of lycopene as a potent antioxidant, it has been the subject of several human intervention studies investigating its beneficial role in male infertility. The evidence does suggests that consuming lycopene as a supplement or contained in foods such as tomatoes and tomato juice improves sperm concentration, motility, and functionality. However, further studies are needed to more precisely define the levels of lycopene to be consumed before dietary recommendations can be made to treat and manage male infertility.

8.5 SKIN DISEASES

Human skin, which is the outer covering of the body, is the largest organ, providing protection for the human body against environmental factors. Its functions include providing immunity defence, insulation, regulation of temperature and water, and synthesis of vitamin D. Several environmental and dietary factors can influence the functionality and health status of the skin. One common factor is exposure to UV rays from the sun. Over the last several years, scientists have been interested in investigating the relationship between diet, nutrients, and skin health. More recently, oxidative stress has been identified as one of the more important factors in the disruption of the normal functioning of the skin, as well as skin damage and diseases. Antioxidants administered either systemic or topical have been suggested as an effective strategy in protecting the skin against UV light-induced photodamage. In this context, both antioxidant nutrients as well as phytochemicals such as carotenoids and polyphenols have been the subject of several investigations over the past decade. Several review articles have been published summarizing the results from these earlier studies [43–47]. Lipid soluble carotenoids, due to their potent antioxidant properties, have been studied extensively. Lycopene and β-carotene are two of the most predominant carotenoids present in fruit and vegetables as well as in the body. A significant correlation between dermal and plasma lycopene as well as β-carotene was observed in healthy human subjects. Low-lycopene diets resulted in lower skin levels that were restored by lycopene supplementation.

Erythema—redness of the skin caused by exposure to UV-light and used as an indication of photodamage of the skin was shown to decrease with lycopene supplementation for a period of 10–12 weeks [43,44]. The protective activity of a dermal product based on lycopene and another one containing a mixture of vitamins E and C were tested [48]. Results showed that the lycopene-based product had a much greater protective ability than the product containing the mixture of vitamins. The authors of the study concluded by suggesting that lycopene use may be effective in the prevention of cutaneous damage by free radicals. A study by Rizwan et al. [49] showed that daily consumption of tomato paste containing 16 mg of lycopene for 12 weeks provides significant protection against acute and perhaps longer-term aspects of photodamage.

Contrary to the observations reported in the literature that lycopene has significant photo-protective effect against UV light, a study by Sokoloski et al. [50] showed that lycopene, either natural or in the form of a pill, had no effect on systemic photoprotection against UVB.

To test the hypothesis that high levels of antioxidant substances may be correlated to lower levels of skin roughness, Maxim et al. [51] used optical non-invasive *in vivo* quantitative methods for measuring the structures of the furrows and wrinkles on the forehead skin, as well as the concentration of lycopene on the forehead skin of 20 subjects. A significant correlation was obtained between the skin roughness and the lycopene concentration.

Polymorphic Light Eruption (PLE) is another similar skin disorder associated with exposure to UV light. In a recent study, a nutritional supplement containing lycopene, β-carotene, and Lactobacillus johnsonii was used to supplement PLE patients to investigate its protective effect against UVA-induced skin lesions. A significant reduction in skin lesions was observed in the nutritional supplemented group compared to controls [52].

In a recent study, Grether-Beck et al. [53] showed that a lycopene-rich tomato nutrient complex and the carotenoid lutein protected against UVA/B- and UVA1-induced gene expression in human skin. Based on this observation they suggest measuring the molecular markers for UVA1-inducible genes to investigate photodamage *in vivo*.

In summary, the emphasis of many studies over the past two decades has been to study the beneficial role of antioxidants in human skin. Several reviews have been published summarizing the results of these studies. Overall, a majority of human intervention studies have shown that lycopene is effective in preventing skin against UVA and UVB damage. However, some reported studies failed to substantiate such beneficial effects of lycopene. Future studies using larger and better-defined subjects, more recent analytical methods to measure circulating skin lycopene levels, and recent methods of measuring skin damage will surely contribute important information that will go a long way in our understanding of the role of antioxidants such as lycopene in skin health. Nutritional strategies can then be formulated for the use of health professionals in protecting the skin from UV light-induced damage, and in improving the quality of life.

8.6   NEURODEGENERATIVE DISEASES (NDD)

Degenerative diseases of the nervous system include Alzheimer's disease (AD), Parkinson's disease, Huntington's diseases, epilepsy, and amyotropic lateral sclerosis (ALS) [54]. Although they vary in their etilogies, oxidative stress is now recognized as one of the most important causative factors and an ancillary factor in the pathogenesis of NDD [55].

This association is based on several observations (Table 8.2). As in other human diseases discussed in this review, antioxidants may have a critical role in mitigating the damaging effects of ROS and protect the brain against diseases. Over the past decades research has been directed towards the role of several antioxidants, including several free radical-deactivating enzymes, vitamin, and mineral agents that scavenge free radicals and phytonutrients [56–59]. A higher daily intake of fruit and vegetables that contain several antioxidant compounds, including carotenoids such as lycopene, has been shown to improve the antioxidant status of healthy individuals and improve their cognitive function, suggesting a relationship between diet and brain function [60,61]. Patients with NDD were shown to have high levels of lipid peroxidation and DNA oxidation and reduced activities of the ROS scavenging enzymes compared to controls [54]. *In vitro* studies using tissue from the nervous system have also demonstrated the protective effect of antioxidants against neuronal damage. Although amongst the carotenoids β-carotene has been studied extensively, studies with lycopene are limited mostly to *in vitro* and animal models. There is a lack of literature relating specifically to the role of lycopene in preclinical and clinical trials as well as epidemiological investigations.

In a recent study [62], the relationship between serum carotenoid levels including lycopene and the risk of Alzheimer's disease-related mortality in a representative population of American adults was investigated. A total of 6,958 subjects older than 50 years participated in the study. Results showed that high serum levels of lycopene and lutein+zeaxanthin at the baseline were associated with a lower risk of AD mortality rates decreased with increasing serum lycopene levels. No association with mortality was found with other carotenoids, including β-carotene.

TABLE 8.2
Vulnerability of Brain to Free Radical Damage

Brain consumes large quantities of oxygen for its relatively small weight, contributing to the formation of ROS.

Membrane lipids in brain contain high levels of polyunsaturated fatty acid side chains that are prone to free radical attack.

Presence of iron and other transition metals in the brain can also contribute to the production of ROS.

Brain contains lower levels of antioxidant vitamins such as vitamins A, C, and E.

Brain contains low to moderate amounts of antioxidant enzymes, such as atalase, superoxide dismutase, and glutathione peroxidase, which play an important role in the inactivation of ROS.

Peripheral levels and activities of a broad spectrum of non-enzymatic and enzymatic antioxidants in elderly subjects with mild cognitive impairment (MCI) and AD were compared against matched controls. Lycopene was one of the many antioxidants estimated. Results showed similar lower peripheral levels and activities of antioxidants including lycopene in MCI and AD patients when compared to controls. It was suggested that antioxidants may be important in preventing the progression of early neuronal damage to more advanced stages such as AD [63].

In a study by Polidori et al. [64], AD patients and vascular AD dementia patients were compared with respect to their plasma water-soluble and lipid-soluble micronutrients as well as levels of biomarkers of lipid and of protein oxidation. With the exception of β-carotene, all other antioxidants—including lycopene—were found to be lower in demented patients as compared to controls. Authors conclude that dementia, immaterial of its nature, is associated with the depletion of a large spectrum of antioxidant micronutrients and with increased protein oxidative modification. However, a study from Germany [65] evaluating the serum levels of antioxidants in persons with mild dementia found no association between plasma lycopene and other nutrient antioxidants and dementia.

A combination of polyunsaturated fatty acids, lycopene, and Ginkgo biloba extracts were administered daily for three years to a group of 41 elderly subjects aged 65 years or older and compared with a control group who did not ingest the supplement. Changes in cognitive function during a three-year follow-up were measured in both groups. Administering the antioxidant combination mixture resulted in an improvement in the cognitive function of patients after three years. It was concluded that administering a mixture of multiple antioxidants acted synergistically in promoting the antioxidant effect and contributed to the improvement of the cognitive function [66].

In summary, since the human brain consumes large quantities of oxygen for its function and contains large quantities of polyunsaturated fatty acids in the lipids of the membrane, it is highly prone to oxidative damage. Antioxidants can play a critical role in preventing the generation and accumulation of ROS and thereby play a beneficial role in protecting the brain from damage. Most of the studies, with the exception of a few, have shown that lycopene, due to its potent antioxidant properties, can prevent oxidative stress and provide protection to the brain. Needless to say, so far only a few human intervention studies have been undertaken in this regard, using a relatively small number of subjects. Further studies are needed to confirm the observed results in the future and also to establish the most effective level of lycopene to be consumed by including sources such as tomatoes and tomato products in the diet. Research is also needed to substantiate the role and level of lycopene in nutritional supplements in brain health.

8.7 RHEUMATOID ARTHRITIS (RA)

Rheumatoid arthritis (RA) is a long-term autoimmune-mediated inflammatory disorder that primarily affects the joints and causes pain. Both genetic and environmental factors including diet are believed to be involved in the causation of RA. Oxidative stress has been shown to closely relate to the development of RA. Based on earlier *in vitro* studies, more recently human studies have been initiated to investigate the role of antioxidants including lycopene in this important human health disorder. A few review articles have been published in this regard [67,68].

In one study, the effect of antioxidant supplements on clinical outcomes and antioxidant parameters in RA patients was investigated [69]. An antioxidant mixture containing zinc and vitamins A, C, and E, but not lycopene, were administered to female RA patients for 12 weeks. Based on the observed results, authors conclude that antioxidants may improve RA activity significantly. However, in this study the number of painful and swollen joints and increased erythrocyte antioxidant levels were not observed. On the other hand, Hagfors et al. showed [70] no correlation between antioxidant intake and the plasma levels of antioxidants including lycopene and other variables related to RA disease activity. However, in another study by Kratas et al. [71] an increased oxidative stress and a low antioxidant status in patients with RA was observed. Similarly, in another study, significantly lower plasma carotenoid levels including lycopene were observed in RA patients compared with non-RA subjects after adjustment for potential confounders [72]. Costanbader et al. [73] did not find any association between antioxidant intake and the risk of developing RA. In an open pilot study the clinical parameters for RA were measured to test the beneficial effects of antioxidant intervention. Although, in general, the laboratory measures of inflammatory activity and oxidative modification were unchanged, the number of swollen and painful joints were significantly decreased and general health significantly increased [74].

In a more recent study, Han and Han studied the role of lycopene in reducing the risk of mortality in RA patients. Patients were divided into three groups according to their serum lycopene concentration. They were then followed up for mortality over a number of years. After adjusting for demographic and other risk factors, RA participants in the higher-lycopene group had a significantly reduced hazard ratio for all-cause mortality compared to RA participants in the low-lycopene group [75].

In summary, although some investigations showed no correlation between antioxidant intervention and RA, others have provided evidence in support of the relationship between oxidative stress and RA as well as the beneficial role of antioxidants in lowering the risk of RA. Overall, it could be concluded that antioxidants such as lycopene may be associated with RA. However, more research, including intervention studies, is needed to fully validate such a conclusion and enhance our understanding of the mechanisms of action and effective supplemental levels of lycopene and other antioxidants.

8.8   OTHER HUMAN HEALTH DISORDERS

Osteoporosis is an important human health disorder around the globe. Oxidative stress has also been shown to increase the risk of osteoporosis. The role of tomatoes, tomato products, and lycopene in the prevention and management of osteoporosis is covered extensively in another chapter of this book. Research is now beginning to be undertaken to investigate the role of lycopene in other human health disorders listed in Table 8.1 and perhaps others not listed. However, the information based on *in vitro* and animal studies is still in its infancy. Fully randomized, double-blind crossover preclinical and clinical studies using a larger number of patients and their controls are yet to be undertaken. It remains for future studies to provide the evidence for the role of lycopene in these human diseases, mechanisms of action, and the effective dosages.

8.9   SUMMARY AND FUTURE DIRECTIONS

In vitro, animal, preclinical, human intervention, and epidemiological studies over the past several decades have supported the role of oxidative stress caused by the accumulation of ROS as being causally related to many, if not all, human diseases. To combat the damaging effects of oxidative stress, antioxidants—including antioxidant nutrients as well as phytochemicals—have been the subject of several investigations over the past several decades. Carotenoids in general and lycopene in particular as a potent antioxidant have been studied extensively. The initial interest in the beneficial effects of lycopene was in its role in prostate cancer. However, this interest expanded to other cancers and CVD that are covered extensively in other chapters of this book. Since oxidative damage to DNA, proteins, and lipids increases the risk of most human diseases, research relating to lycopene in diseases other than cancer and CVD soon began. Systematic *in vitro* and animal studies as well as preclinical, clinical, and epidemiological studies have been done with relation to some diseases, such as the ones reviewed in this chapter. Although there are some contradictory observations reported in the literature, overall antioxidants such as lycopene are shown to increase the antioxidant capacity in the body and prevent the risk of these diseases. However, the exact mechanisms of action and the effective dose of the antioxidants have not yet been established. Studies are now underway to address this aspect of human health. Studies investigating the role of lycopene in other diseases such as rhumatoid arthritis and oral and inflammatory disorders are still in their infancy. It is important for any future studies in these areas to give particular attention to details of study protocols that include fully randomized and double-blind procedures as well as well-defined patients with respect to the nature of the disease, age, and gender, proper controls, larger size of subjects, longer periods of intervention, use of physiologically relevant lycopene dose levels, and the measurement of analytical, biochemical and clinical end points that are associated with antioxidant properties and the disease. Only then can we make definitive conclusions and provide effective recommendation as to the use of lycopene and other antioxidants in the prevention, management, and perhaps treatment of human diseases.

LIST OF ABBREVIATIONS

ACOS   asthma-COPD overlap syndrome
BA   bronchial asthma
COPD   chronic obstructive pulmonary disease
CVD   coronary vascular disease
DASH   dietary approaches to stop hypertension
IVF   invitrofertilisation
MCI   mild cognitive impairment

NDD	neurodegenerative diseases
NO	nitrous oxide
PME	polymorphic light eruption
RA	rheumatoid arthritis
ROS	reactive oxygen species

REFERENCES

1. Giovannucci, E., E. B. Rimm, Y. Liu, M. J. Stampfer, W. C. Willett. 2002. "A prospective study of tomato products, lycopene, and prostate cancer risk." *J Natl Cancer Inst* 94(5):391–8.
2. Public Health Agency of Canada. 2014. https://www.canada.ca/en/public-health/services/chronic-diseases/chronic-respiratory-diseases.html . (Date modified: December 16, 2014).
3. Shanmugasundaram, K. R., S. S. Kumar, S. Rajajee. 2001. "Excessive free radical generation in the blood of children suffering from asthma." *Clin Chim Acta* 305:107–14
4. Nadeem A., S. K. Chhabra, A. Masood, H. G. Raj. 2003. "Increased oxidative stress and altered levels of antioxidants in asthma." *J Allergy Clin Immunol* 111:72–8.
5. Hosseini B., B. S. Berthon, P. Wark, L. C. Wood. 2017. "Effects of fruit and vegetable consumption on risk of asthma, wheezing and immune responses: A systematic review and meta-analysis." *Nutrients* 29:9.
6. Wood L. G., M. L.Garg, J. M. Smart, H. A., D. Barker, P. G. Gibson. 2012. "Manipulating antioxidant intake in asthma: A randomized controlled trial." *Am J Clin Nutr* 96:534–43.
7. Wood L. G., M. L. Garg, H. Powell, P. G. Gibson. 2008. "Lycopene-rich treatments modify noneosinophilic airway inflammation in asthma: Proof of concept." *Free Radic Res* 42: 94–102.
8. Wood, L. G., M. L. Garg, R. J. Blake, S. Garcia-Caraballo, P. G. Gibson. 2005. "Airway and circulating levels of carotenoids in asthma and healthy controls." *J Am Coll J Am Coll Nutr* 24:448–55.
9. Kodama Y., Y. Kishimoto, N. Muramatsu, J. Tatebe, Y. Yamamoto, N. Hirota, Y. Itoigawa, R. Atsuta, K. Koike, T. Sato, K. Aizawa, K. Takahashi, T. Morita, S. Homma, K. Seyama, A. Ishigami. 2016. "Antioxidant nutrients in plasma of Japanese patients with chronic obstructive pulmonary disease, asthma- COPD over-lap syndrome and bronchial asthma." *Clin Respir J* 1–10.
10. Kasagi S., K. Seyama, H. Mori, S. Souma, T. Sato, T. Akiyoshi, H. Suganuma, Y. Fukuchi. 2006. "Tomato juice prevents senescence-accelerated mouse P1 strain from developing emphysema induced by chronic exposure to tobacco smoke." *Am J Physiol Lung Cell Mol Physiol* 290:396–404.
11. Rao A. V., L. G. Rao, Ray, M. R. 2006. "Lycopene." *Adv Food and Nutri Res* 51:99–164.
12. Friedman J., E. Peleg, T. Kagan, S. Shnizer, T. Rosenthal. 2003. "Oxidative stress in hypertensive, diabetic, and diabetic hypertensive rats." *Am J Hypertens* 16:1049–52.
13. Lassegue B. and K. K. Griendling. 2004. "Reactive oxygen species in hypertension." *Am J Hypertens* 17:852–60.
14. Zaidi, M., A. S. M. T. Alam, C. M. R. Bax, B. E. Bax, V. S. Shankar, J. S. Gill, M. Pazianas, C. L. H. Huang, T. Sahinoglu, B. S. Moonga, C. R. Stevens, D. R. Blake. 1993. "Role of the endothelial cell in osteoclast control: New perspectives." *Bone* 14:97–102.
15. Most M. M. 2004. "Estimated phytochemical content of the dietary approaches to stop hypertension (DASH) diet is higher than in the control study diet." *J Am Diet Assoc* 104:1725–7.
16. Rocchi E., A. Borghi, F. Paolillo, M. Pradelli, G. Casalgrandi. 1991. "Carotenoids and liposoluble vitamins in liver cirrhosis." *J Lab Clin Med* 118:176-18.

17. Paran E. and Y. Engelhard. 2001. "Effect of Lyc-O-Mato, standardized tomato extract on blood pressure, serum lipoproteins, plasma homocysteine and oxidative stress markers in grade 1 hypertensive patients." Proceedings of the 16th Annual Scientific Meeting of the Society of Hypertension, San Francisco, USA.
18. Rao, A. V. and L. G. Rao. 2007. "Carotenoids and human health." *Pharmacol Res* 55:207–16.
19. Paran E. 2006. *Reducing hypertension with tomato lycopene. In: Rao AV, editor, Tomatoes, lycopene and human health.* Scotland: Caledonian Science Press; p.169–82.
20. Engelhard Y. N., B. Gazer, E. Paran. 2006. "Natural antioxidants from tomato extract reduce blood pressure in patients with grade-1 hypertension: A double-blind, placebo-controlled pilot study." *Am Heart J* 151:100.e1–100.e6.
21. Paran E., V. Novack, Y. N. Engelhard, N. Yechiel, I. Hazan-Halevy. 2009. "The effects of natural antioxidants from tomato extract in treated but uncontrolled hypertensive patients." *Cardiovasc Drugs Ther* 23:145–51.
22. Moriel, P., A. Sevanian, S. Ajzen, M. T. Zanella, F. L. Plavnik, H. Rubbo, D. S. P. Abdalla. 2002. "Nitric oxide, cholesterol oxides and endothelium-dependent vasodilation in plasma of patients with essential hypertension." *Med Biol Res* 35:1301–9.
23. Subhash K., C. Bose, B. K. Agrawal. 2007. "Effect of lycopene from tomatoes (Cooked) on plasma antioxidant enzymes, lipid peroxidation rate and lipid profile in Grade-I hypertension." *Ann Nutr Metab* 51:477–81.
24. Xinli L. and X. Jiuhong. 2013. "Lycopene supplement and blood pressure: An updated meta-analysis of intervention trials." *Nutrients* 5:3696–712.
25. Hozawa A., D. R. Jacobs, Jr, M. W. Steffes. 2009. "Circulating carotenoid concentrations and incident hypertension: The Coronary Artery Risk Development in Young Adults (CARDIA) study." *J Hypertens* 27:237–42.
26. Yu T. X. Y., X. Dong, P. YaFei, L. JunHua. 2009. "Protective effects of lycopene against H2O2-induced oxidative injury and apoptosis in human endothelial cells." *Cardiovasc Drugs Ther* 23:439–48.
27. Hermann M., A. Flamme, T. F. J. Lüscher. 2006. "Nitric oxide in hypertension." *J Clin Hypertens* (Greenwich) Suppl 4:17–29.
28. Dalbeni A., D. Treggiari, B. Fava, C. Pandolfini, T. M. Pietro. 2016. "Lycopene increases nitric oxide bioavailability and inhibits endothelial cells migration." *J Hypertension* 34(suppl 2):e24.
29. Dubin L. and R. D. Amelar. 1971. "Etiologic factors in 1294 consecutive cases of male infertility." *Fertil Steril* 22:469–74.
30. Greenberg, S. H., L. L. Lipshuitz, A. J. Wein. 1978. "Experience with 425 subfertile male patients." *J Urol* 119:507–10.
31. Johnson, W. 1975. "Proceedings. 120 Infertile men." *Br J Urol* 47:230.
32. Askin H., Z. Kurcer, F. A. Faruk, F. Baba, S. Engin, A. Ahmet. 2009. "Lycopene, an antioxidant carotenoid, attenuates testicular injury caused by ischemia/reperfusion in rats." *Tohoku J Exp Med* 218:141–7.
33. Agarwal A., C. Ong, P. Prashast. 2014. "Lycopene and male infertility." *Damayanthi Durairajanayagam, Asian J Androl* 16:420–32.
34. Iwasaki A. and C. Gagnon. 1992. "Formation of reactive oxygen species in spermatozoa of infertile patients." *Fertil Steril* 57:409–16.
35. Pakrashi T. and S. Oehninger 2014. "Lycopene and male infertility: Do we know enough?" *Asian J Androl* 16:500.
36. Moncada M. L., E. Vicari, C. Cimino, A. E. Calogero, A. Mongioi, R. D'Agata. 1992. "Effect of acetylcarnitine in oligoasthenospermic patients." *Acta Eur Fertil* 23:221–4.
37. Palan P. and R. Naz. 1996. "Changes in various antioxidant levels in human seminal plasma related to immunofertility." *Arch Androl* 36:139–43.

38. Mohanty N. K., R. Kumar, N. P. Gupta. 2001. "Lycopene therapy in the management of idiopathic oligoasthenospermia." *Ind J Urol* 56:102–103.

39. Durairajanayagam D., A. Agarwal, C. Ong, P. Prashast. 2014. "Lycopene and male infertility." *Asian J Androl* 16:420–32.

40. Faridullah H., T. Eva, M. Peter, R. Stawarz, L. Norbert. 2015. "Effects of biological active substances to the spermatozoa quality." *The Journal of Microbiology Biotechnology and Food Sciences* 5:263–7.

41. Filipcikova R., I. Oborna, J. Brezinova. 1213. "Lycopene improves the distorted ratio between AA/DHA in the seminal plasma of infertile males and increases the likelihood of successful pregnancy." *Biomedical papers of the Medical Faculty of the University Palacký, Olomouc, Czechoslovakia* 159:77.

42. Yamamoto, Y. K. Aizawa, M. Mieno, M. Karamatsu, Y. Hirano, K. Furui, T. Miyashita, K. Yamazaki, T. Inakuma, I. Sato, H. Suganuma, T. Iwamoto. 2017. The effects of tomato juice on male infertility. *Asia Pac J Clin Nutr* 26:65–71.

43. Wilhelm S. and S. Helmut. 2007. "Carotenoids and Flavonoids Contribute to Nutritional Protection against Skin Damage from Sunlight." *Mol Biotechnol* 37:26–30.

44. Stall W. H. 2012. "Photoprotection by dietary carotenoids: Concept, mechanisms, evidence and future development." *Mol Nutr Food Res* 56:287–95.

45. Juergen L., M. C. Martina, S. Wolfram, D. Maxim. 2011. "Carotenoids in human skin." *Clin Exp Dermatol* 20:377–82.

46. Jalil P. I. and M. Habib. 2015. "Lycopene as a carotenoid provides radioprotectant and antioxidant effects by quenching radiation-induced free radical singlet oxygen: An overview." *Cell* 16:386-391.

47. Fernández-García E. 2014. "Skin protection against UV light by dietary antioxidants." *Food Funct* 5:1994–23.

48. Andreassi M., E. Stanghellini, A. Ettorre, A. Di Stefano, L. Andreassi. 2004. "Antioxidant activity of topically applied lycopene." *J Eur Acad Dermatol Venereol* 18:52–5.

49. Rizwan M., I. Rodriguez-Blanco, A. Harbottle, M. A. Birch-Machin, R. E. B. Watson, L. E. Rhodes. 2011. "Tomato paste rich in lycopene protects against cutaneous photodamage in humans in vivo: A randomized controlled trial." *Br J Dermatol* 164:154–62.

50. Sokoloski L., M. Borges, E. Bagatin. 2015. "Lycopene not in pill, nor in natura has photoprotective systemic effect." *Arch Dermatol Res* 307:545 –9.

51. Maxim D., A. Patzelt, S. Gehse et al. 2008. "Cutaneous concentration of lycopene correlates significantly with the roughness of the skin." *Eur J Pharm Biopharm* 69:943–7.

52. Marini, A., T. Jaenicke, S. Grether-Beck, C. Le Floc'h, A. Cheniti, N. Piccardi, J. Krutmann. 2014. "Prevention of polymorphic light eruption by oral administration of a nutritional supplement containing lycopene, β-carotene, and Lactobacillus johnsonii: Results from a randomized, placebo-controlled, double-blinded study." *Photodermatol Photoimmunol Photomed* 30:89–194.

53. Grether-Beck S., A. Marini, T. Jaenicke, W. Stahl, J. Krutmann. 2017. "Molecular evidence that oral supplementation with lycopene or lutein protects human skin against ultraviolet radiation: Results from a double-blinded, placebo-controlled, crossover study." *Br J Dermatol* 176:1231–40.

54. Rao A. V. and B. Balachandran. 2003. "Role of oxidative stress and antioxidants in neurodegenerative diseases." *Nutr Neurosci* 5:291–309.

55. Mariani E., M. C. Polidori, A. Cherubini, P. Mecocci. 2005. "Oxidative stress in brain aging, neurodegenerative and vascular diseases: An overview." *J Chromatogr* 827:65–75.

56. Singh R. P., S. Sharad, S. K. Singh. 2004. "Free radicals and oxidative stress in neurodegenerative diseases: Relevance of dietary antioxidants." *J Indian Acad Clin Med* 5:218–25.

57. Takeda A., O. P. Nyssen, A. Syed, E. Jansen, B. Bueno-de-Mesquita, V. Gallo. 2013. "Vitamin A and carotenoids and the risk of Parkinson's disease: A systematic review and meta-analysis." *Neuroepidemiology* 42:25–38.
58. Obulesu M., D. M. Rao, P. V. Bramhachari. 2011. "Carotenoids and Alzheimer's disease: An insight into therapeutic role of retinoids in animal models." *Neurochem Int* 59:535–41.
59. Charan D. C. and P. Poonam. 2014. "The discovery and development of new potential antioxidant agents for the treatment of neurodegenerative diseases." *Expert Opin Drug Discov* 9:1205–22.
60. Polidori M. C., D. Praticó, F. Mangialasche, E. Mariani, O. Aust, T. Anlasik, N. Mang, L. Pientka, W. Stahl, H. Sies, P. Mecocci, G. Nelles. 2009. "High fruit and vegetable intake is positively correlated with antioxidant status and cognitive performance in healthy subjects." *J Alzheimers Dis* 17:921–7.
61. Brenner S. R. 2017. "Alzheimer's disease and other neurodegenerative diseases may be due to nutritional deficiencies secondary to unrecognized exocrine pancreatic insufficiency." *Med Hypotheses* 102:89–90.
62. Min J.-Y. and K.-B. Min. 2014. "Serum lycopene, lutein and zeaxanthin, and the risk of Alzheimer's disease mortality in older adults." *Dement Geriatr Cogn Disord* 37:246–56.
63. Rinaldi P., M. C. Polidori, A. Metastasio et al. 2003. "Plasma antioxidants are similarly depleted in mild cognitive impairment and in Alzheimer's disease." *Neurobiol Aging* 24:915–19.
64. Polidori, M. C., P. Mattioli, S. Aldred et al. 2004. "Plasma antioxidant status, immunoglobulin G oxidation and lipid peroxidation in demented patients: Relevance to Alzheimer disease and vascular dementia." *Dement Geriatr Cogn Disord* 18:265–70.
65. von Arnim, C. A. F., F. Herbolsheimer, T. Nikolaus, R. Peter, H. K. Biesalski, A. C. Ludolph, M. Riepe, N. Matthias, G. Nagel. 2012. "Dietary antioxidants and dementia in a population-based case-control study among older people in South Germany." *J Alzheimers Dis* 31:717–24.
66. Yasuno F., S. Tanimukai, M. Sasaki, C. Ikejima, F. Yamashita, C. Kodama, K. Mizukami, T. Asada. 2012. "Combination of antioxidant supplements improved cognitive function in the elderly." *J Alzheimers Dis* 32:895–903.
67. Giuseppe, D. D. and A. Wolk. 2014. "Diet and rheumatoid arthritis development: What does the evidence say?" *Int J Clin Rheumtol* 9:169–82.
68. Al-Okbi S. Y. 2014. "Nutraceuticals of anti-inflammatory activity as complementary therapy for rheumatoid arthritis." *Toxicol Ind Health* 30:738–49.
69. Jalili M., S. Kolahi, R.-F. Aref-Hosseini, M. E. Mamegani, A. Hekmatdoost. 2014. "Beneficial role of antioxidants on clinical outcomes and erythrocyte antioxidant parameters in rheumatoid arthritis patients." *Int J Prev Med* 5:835–40.
70. Hagfors, L., P. Leanderson, L. Sköldstam, J. Andersson, G. Johansson. 2003. "Antioxidant intake, plasma antioxidants and oxidative stress in a randomized, controlled, parallel, Mediterranean dietary intervention study on patients with rheumatoid arthritis. *Nutr J* 2:5–12.
71. Karatas F., I. Ozates, H. Canatan, I. Halifeoglu, M. Karatepe, R. Colakt. 2003. "Antioxidant status & lipid peroxidation in patients with rheumatoid arthritis." *Indian J Med Res* 118:178–81.

72. De Pablo P., T. Dietrich, E. W. Karlson. 2007. "Antioxidants and other novelcardiovascular risk factors in subjects with rheumatoid arthritis in a large population sample. *Arthritis Rheumatol* 57:953–62.
73. Costenbader K. H., J. H. Kang, E. W. Karlson. 2010. "Antioxidant intake and risks of rheumatoid arthritis and systemic lupus erythematosus in women." *Am. J. Epidemiol* 172:205–16.
74. van Vugt R. M., P. J. Rijken, A. G. Rietveld, A. C. van Vugt, B. A. C. Dijkmans. 2008. "Antioxidant intervention in rheumatoid arthritis: Results of an open pilot study." *J Clin Rheumatol* 27:771–5.
75. Han G.-M. and X.-F. Han. 2016. "Serum lycopene is inversely associated with long-term all-cause mortality in individuals with rheumatoid arthritis: Result from the NHANES III." *Eur J Integr Med* 8:213–18.

9 Lycopene and Tomatoes

Luca Sandei

CONTENTS

9.1   INTRODUCTION

Let Food Be Your Medicine, and Let Medicine Be Your Food.

Hippocrates, 400 BC

You are what you eat.

Feuerbach, 1885

Studies and reports from across the globe have well established the link between dietary patterns, lifestyle habits, and chronic illnesses, and have served as the basis for developing dietary guidelines. Many aspects of our daily life have changed over the last few years. The world is smaller and wholly interconnected, life is more frenetic, and foods will have to evolve progressively towards new emergent attitudes, not only to be used as simple nutrition-releasers, but as key players for the human well-being, for sustainable agriculture management, and as new health performers for modern society's demands. More than 100 illnesses in humans, from cancer to cardiovascular diseases, are caused by oxidative stresses due to reactive oxygen species involving biological macromolecules [1] of radical type and by free radicals.

In extremely simple terms, it can be said that a free radical is an extremely reactive and short-life molecule, which has one or more unpaired electrons and which can damage the biological molecules near it.

9.2 RADICALS AND NATURAL ANTIOXIDANTS

Free radicals usually form in the organism, but their development can be also caused by environmental conditions and certain behaviours (atmospheric pollution, smoke, alcohol, exposure to radiation, etc.) that cause the formation of other reactive species with toxic effects [2]. Radical reactions mainly involve lipids in peroxidation processes, from which other cellular damages derive. Important factors, able to influence the oxidation reactions, which may occur in the different substrates, are pH value, temperature, the unsaturation degree of the fatty acid present in the lipid matrix (triacylglycerols, phospholipids), the presence of transition-metal ions and oxygen availability. Antioxidants act either as electron donors or as free-radical scavengers, able to decompose peroxides, inhibit singlet oxygen, and to chelate some metals. Initiation and propagation of *in vivo* radical reactions follow very complex pathways, for which reference is made to specialist texts. Vitamins and antioxidant pigments can eliminate free radicals, thus reducing oxidative stress and its resulting symptoms.

A modern Mediterranean diet, owing to the abundance of fresh vegetables, supplies a considerable amount of micro and macro components that play an important role as antioxidants protecting the cells of various tissues from oxidative damage due to both internal and external factors [3].

Indeed, more people are recognizing the value of "close-to-nature" products, as evidenced by their increased reliance on nature-based antioxidant foods and supplements. One of the most important trends in the current food industry is "total sustainability" behavior, with consequent growing demand for "green food" obtained only with environmental care, crop management evolution, soil fertility attention, and, above all, the creation of novel powerful healthy fresh foods and novel "naturally enriched" specific functional processed foods.

Mother Nature is a fantastic intrinsic synthesizer of flavor, fragrances, and bioactive compounds in almost all its natural products, especially in fruit and vegetables [4]. Vegetables contain many bioactive compounds that contribute to supporting the improvement of humans' health, helping to decrease the risk of many "modern" diseases (cancer, cardiovascular and neurological diseases, macular degeneration, cataracts, osteoporosis, hypertension, etc.) [5]. A long life depends on many biological strategies to promote cell survival against many internal and external stimuli [6]. Physical functionality, biological attributes, psychological factors, and social environment and support are the main determinants of adequate nutrition and functional and neurophysiological capacities, especially in aging people [7].

The nutritional relevance of fruit and vegetables on human health can be attributed to their dietary fiber content, mainly soluble, as well as to a large range of micronutrients, including carotenoids, vitamins (mainly vitamin C and folate), and minerals (such as potassium, calcium, and magnesium). In addition, fruit and vegetables are well-recognized sources of non-nutrient bioactive compounds, also called "phytochemicals," among which (poly)phenolic compounds are the predominant and the most investigated [8–13]. The term (poly)phenol includes a number of different chemical structures, including flavonoids and related compounds, but also refers to hydroxycinnamates and phenolic acids, which have only one phenolic ring.

Many substances naturally occurring in vegetables have the ability to react with free radicals, breaking the chain reactions that lead to the formation of more radicals or acting as scavengers of ROS (oxygen reactive species). To protect itself against an excessive production of free radicals, the human body has developed sophisticated mechanisms to maintain redox homeostasis, increasing its removal or blocking its production through endogenous antioxidant defences, either enzymatic or non-enzymatic, which are accompanied by exogenous antioxidant defences, mostly represented by the antioxidants in the diet.

Formerly considered only an ornamental plant, the tomato (*Solanum lycopersicum Mill S.*) has become one of the most commonly consumed vegetable crops, either fresh or processed, and contains recognized bioactive compounds [14].

Currently, processed tomatoes are one of the most important worldwide food commodities. According to the latest figures released by WPTC (World Processing Tomato Council), the worldwide total quantity processed amounted to about 38 million metric tonnes in 2016 (Figure 9.1). The quantity processed in the AMITOM region (Mediterranean Area Processing Association) represented 16.03 million tons; NAFTA countries processed 12.44 million tons (of which 11.47 million were processed in California). Among the major processing countries, Italy processed 5.18 million tons in the last season, split equally across the two Distretto areas (2.84 million tons in the north and 2.77 million tons in the south) [15].

The long history of the processing tomato industry explicates the richness and diversity, which are based on processing chains, of deeply differentiated procedures and on novel technologies that are increasingly refined with the use, for instance, of recycling procedures [16].

The finished products that are available in the different distribution channels have mainly undergone two main processing stages: in the first processing stages fresh tomatoes, as an agricultural raw material coming from direct harvesting operations in the field, are processed into finished and semi-finished intermediary

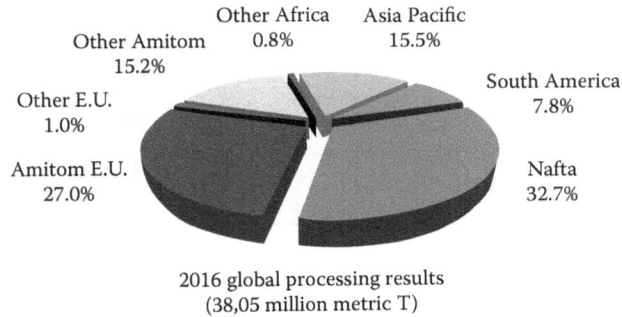

2016 global processing results
(38,05 million metric T)

FIGURE 9.1   World processed tomatoes in 2016.

industrial products (tomato juice, purées, paste, crushed, diced, whole peeled, and unpeeled). In a second processing stage (reprocessing) the intermediate industrial products, conditioned in industrial-format packaging (200-liter drums, bins, etc.), were turned into finished products like tomato sauces, soups, tomato ketchups, pizza toppings, and other formulated products with or without chunks, or diced (Figure 9.2).

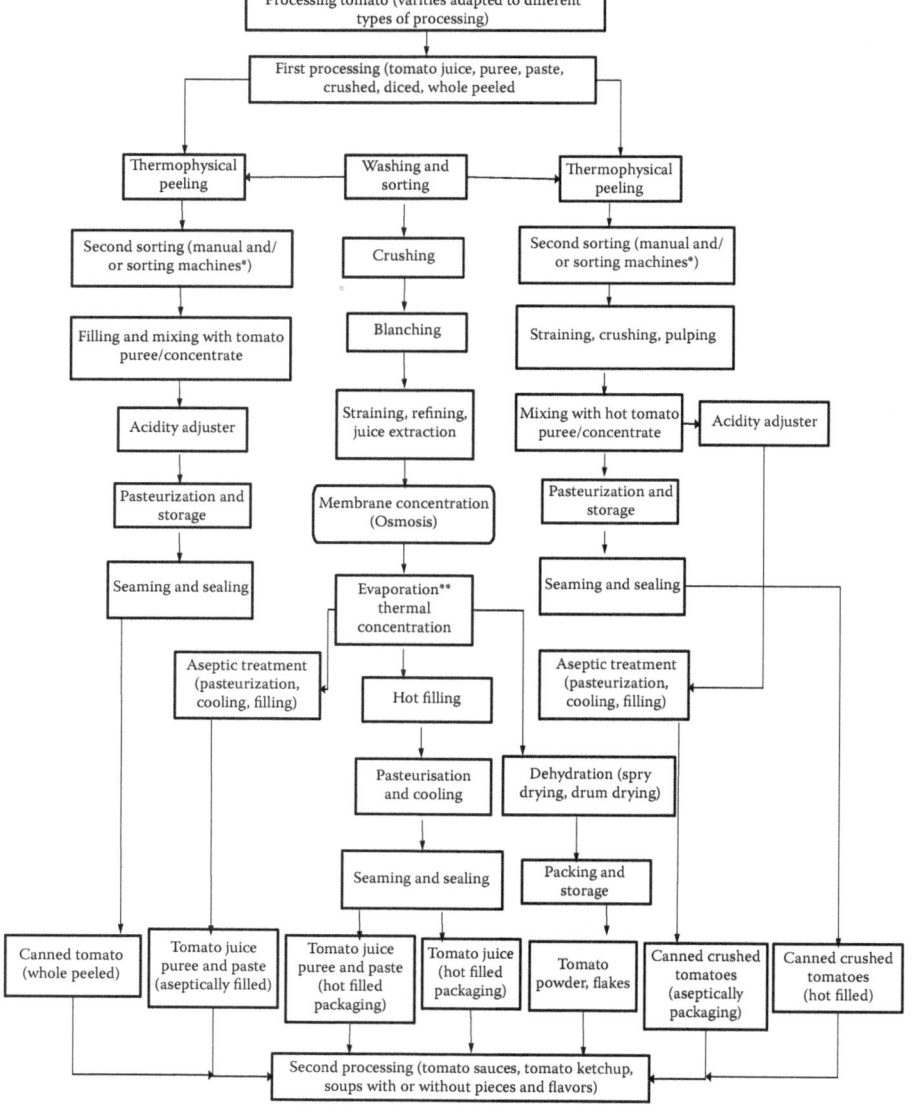

FIGURE 9.2   Main unitary operations of the industrial tomato derivative process.

Tomatoes and tomato products make up a huge part of the worldwide daily vegetable intake. Processed tomatoes, selected and processed at the top of the tomato-harvesting season, including canned tomatoes (whole, diced, etc.), tomato juice, tomato paste, tomato soup, tomato sauce, and ketchup, can contribute to providing the health benefits of eating more vegetable products every day.

Humble but full of precious virtues, the tomato has emerged recently as a high-potential base to produce nutraceutical and functional food product (a healthy food that combines selected nutritional components and the therapeutic properties of naturally present active principles of proven and recognized effectiveness in human health) [17].

Scientific research, in order to develop safe and effective new drugs, has recently rediscovered natural substances as a potential reservoir of innovative therapeutic solutions, with the prospect of integrating and sometimes replacing conventional medicines [18].

9.3 TOMATO PRODUCTS: NUTRITIONAL QUALITY AND HEALTH BENEFITS

The health-protective effects of tomatoes and tomato products have been attributed to the intrinsic synergic combination of the fat-soluble carotenoids (lycopene, β-carotene, and others) as well as water-soluble vitamin C, potassium, and compounds of intermediate hydrophobicity, such as phenolic compounds (quercetin glycosides, rutin, naringenin, and kaempferol). All of these are known to contribute significantly to the specific antioxidant activity of tomatoes [19,20].

The main carotenoids of processed tomatoes are lycopene, β-carotene, and lutein. The major phenolic compounds of tomatoes are the flavanones (naringenin glycosylated derivatives) and the flavonols (rutin, quercetin, and kaemferol glycosylated derivatives) [21].

The tomato seems to have a central role in counteracting the pathogenesis of aging and various degenerative diseases such as atherosclerosis, cardiovascular diseases, diabetes, skin damage, cancer, oxidative stress, and the subsequent formation of free radicals [22].

Oxidation reactions, which are part of aerobic cell metabolism, lead to the formation of free radicals—waste produced by catabolism that, over the years, accumulates and plays a potent oxidant role, which is harmful for almost all constituents of the human body. Free radicals, if not inactivated by the body's antioxidant mechanisms, can worsen many pathological processes.

9.4 TOMATO NATURAL BIOACTIVE MOLECULES

The antioxidant activities of lycopene and other carotenoids are highlighted by their singlet oxygen quenching properties and their ability to trap peroxyl radicals. The quenching activity of carotenoids mainly depends on the number of conjugated double bonds and is clearly influenced by carotenoid end groups (cyclic or acyclic) [23].

The intervention of antioxidants leads to the formation of a radical, which does not allow the propagation of chain reactions of reactive radical species—this radical being rather stable and involved in biological redox systems that regenerate the carotenoid's initial structure [24]:

    1) ROO. + CAR   ROOH + CAR.
    2) CAR.   REDOX CAR.

As for singlet oxygen ($^1O_2$), inactivation by bioactive compounds (lycopene and others) involves transferring excitation to the bioactive compounds (CAR), thus obtaining triplet oxygen (ground state) and excited bioactive compound (CAR). This ability of bioactive compounds to deactivate singlet oxygen results from its molecular structure, which consists, in lycopene for example, in a long chain of 11 conjugated double bonds that is devoid of aromatic rings.

The content of antioxidants in tomatoes is relatively high. On average, 1 kg of fresh product supplies 500 mg of various antioxidants: 160–240 mg of vitamin C, 100–200 mg of total carotenoids, 30–50 mg of phenolic acids, about 10 mg of vitamin E, 20–60 mg of flavonoids, and 0.1–0.4 mg of folates, in addition to traces of minerals of a certain importance such as potassium, selenium, copper, and zinc [25].

It has been scientifically proven that consumption of tomatoes and tomato products is inversely related to the development of certain diseases (i.e., cancers and plasma-lipid peroxidation), thanks to the contents of lycopene, β-carotene, vitamins, polyphenols, and other antioxidant substances, which act synergistically [26]. However, recent studies evidenced that it is not possible to establish a direct relation between protective effect and lonely lycopene consumption. In other words, the administration of lycopene and other natural antioxidants as diet integrators is not as effective as tomato product consumption in preventing these diseases. This suggests the presence, in foods, of a complex molecule system (precursors, unidentified antioxidants, synergists), which, through more or less complex mechanisms, contribute to the overall antioxidant properties of the food [27]. In other words, eating healthy food is better than taking pills.

9.5 LYCOPENE

Among the biomolecules of major food interest, lycopene stands out, the carotenoid most present in tomatoes and, above all, in industrial products (its hydrophobic nature makes it very stable in heat treatment), where it represents 85% of the total amount of the carotenoids measured.

Lycopene must be considered above all for its antioxidant properties, which made it among the various carotenoids, that with the strongest ability to inactivate singlet oxygen and to remove free radicals by preventing LDL oxidation. Tomatoes supply a balanced mixture of water-soluble and fat-soluble antioxidants able to protect cell membranes and their contents from attacks by radicals. Antioxidants of vegetable origin display different functions, such as scavenging of free radicals, quenching of singlet oxygen, chelation of metals, and inhibition of enzymes involved in the formation of reactive oxygen species (ROS) [28].

Lycopene content amounts to 85–90% of the total pigments present in ripe tomatoes for processing; phytoene, phytofluene, β-carotene, and other carotenoids are present in an amount decidedly lower than lycopene. According to Schierle et al. [29], lycopene occurs in tomatoes at levels ranging from 6 to 520 mg/kg (considering both varietal differences and distribution in fruits) but the variability normally encountered in processing tomatoes is comprised between 90 and 180 mg/kg [30]. During ripening, chlorophyll and some carotenes undergo a decrease in level down to chlorophyll total disappearance, whereas lycopene increases with an increase in fruit red color as a result. In addition, present but at a concentration markedly lower than that of lycopene is β-carotene, which presents a maximum concentration in periods preceding full maturation of the fruit followed by a decrease; this clearly shows that final ripening of tomatoes involves a conversion of β-carotene to lycopene, via opening of the final ionone ring [31].

During ripening, temperature and exposure to solar radiations influence carotenoid content and therefore fruit color intensity. Lycopene synthesis is possible in the 12–32°C temperature range, but it is inhibited by an excessive insolation, so that the best cultivation conditions are average temperatures and a good foliage density to protect the fruits against direct exposure to sun [32], even though a rather good illumination seems to be important for carotenoid synthesis in general [33]. Lycopene is more sensitive to high temperature than β-carotene.

Carotenoids, in most cases, are terpene compounds formed by eight isoprene units and exist either as unsubstituted carbohydrates (carotenes) or with one or more polar functional groups (xanthophyll). Carotenes are soluble in oil and non-polar organic solvents, whereas, in aqueous media, they tend to aggregate and precipitate as coagula, improperly called crystals. There are also conformational cis-isomers, which are more polar than the corresponding all-trans isomers: they are less prone to crystallization and are slightly more soluble in aqueous media. Cis forms are more common in plasma [34].

9.6   LYCOPENE: BIOAVAILABILITY AND ANTIOXIDANT ACTION

The carotenoids present in highest concentration in human plasma are β-carotene, lycopene, and lutein. Their presence is due to a diet rich in fruit and vegetables; however, a direct relation between carotenoid intake and their concentration in blood has not yet been found. The presence of carotenoids in blood also shows a certain variability, due to the seasonality of the different vegetable productions. Lycopene has been considered the most efficient "scavenger" of singlet oxygen and remarkably restores those gap-junctions (necessary for communication between cells) that get lost during transformation into malignant cells, re-establishing the lost ones and thus preventing this procedure [35].

Carotenoids seem to be absorbed by the small-intestine mucosa, via a passive process of controlled diffusion. Absorption appears to require incorporation into intestinal micelles in mixture with bile salts: beta-carotene and other carotenoids with pro-vitamin A activity are, in part, converted to vitamin A, and in part incorporated into chylomicrons and secreted into lymph in order to be transferred to liver; the not pro-vitamin ones (lycopene and lutein) seem to be absorbed unaltered. In fasting conditions, in plasma hydrocarbon carotenoids are mainly encountered in plasma LDL (low-density lipoproteins), whereas lutein, zeaxanthin, and xanthophylls are distributed between LDL and HDL (high-density lipoproteins), probably located in a hydrophobic central part. Many factors (environmental, dietary, physiologic, and matrix-dependent) may affect carotenoid bioavailability from foods [36].

Absorption efficiency is influenced by the matrix extraction efficiency, by the presence of triacylglycerols in the mass in an amount sufficient to solubilize carotenoids and to stimulate chylomicron synthesis, and by the presence, in the intestinal lumen, of interfering factors such as pectins and other fibers [37].

In the literature, it has been shown that lycopene absorption by the human organism is higher if tomatoes have been previously processed and have undergone physical or enzymatic treatments resulting in cell wall destruction. Indeed, bioactive molecule absorption is greater for heat-treated tomato products as compared with fresh tomatoes or untreated tomato products, in which cell walls have not been destroyed (cold-break treatment shows different behaviour in comparison with hot-break heat treatment). The physical state and the kind of technological treatment, which causes a certain degree of matrix destruction, play a very important role in lycopene reactivity and in its bioavailability [38].

9.7   MANAGEMENT OF BIOACTIVE COMPONENTS IN TOMATOES

The tomato seems to have a central role in counteracting the pathogenesis of aging and various degenerative diseases such as atherosclerosis, cardiovascular diseases, diabetes, cancer, oxidative stress, and the subsequent formation of free radicals. Oxidation reactions, which are part of aerobic cell metabolism, lead to the formation of free radicals, wastes produced by catabolism that, over the years, accumulate and play a potent role in oxidant action, which is harmful for almost all constituents of the human body. Free radicals, if not inactivated by the body's antioxidant mechanisms, can worsen many pathological processes.

Cultivar selection, environmental factors, agronomic procedures, stage of ripeness at harvest, and appropriate handling, processing, and conditioning influence final lycopene content (all the way from the field to the consumer). For example, it has been calculated that if in red ripe processing tomato cultivars 60–180 mg/kg lycopene are present on average, the concentration of lycopene drops to 5–10 mg/kg in green/yellow cultivars. Responsible for the characteristic deep red color, lycopene's high antioxidant activity intervenes on singlet oxygen ($^1O_2$), to which is due the formation of free radicals by bringing it back to its triplet state through a mechanism of de-excitation (quenching) [39]. As the human body does not autonomously synthesize it, lycopene can be taken only through diet and/or food supplements. The bioavailability of lycopene for humans is strongly influenced by the concomitant presence of lipids in the diet, as they have a key role in the extraction of carotenoids from the aqueous phase with the formation of mixed micelles through which carotenoids are absorbed by enterocytes and transferred to the tissues automatically as lipophilic substances [40].

Lycopene is an important carotenoid pigment with formula $C_{40}H_{56}$ featured a symmetrical structure and containing 13 acyclic carbon-carbon double bonds, of which 11 are conjugated [41]. Lycopene is an isoprenoid molecule of vegetable origin and cannot be synthesized in animal organisms. Lycopene and β-carotene have a chain

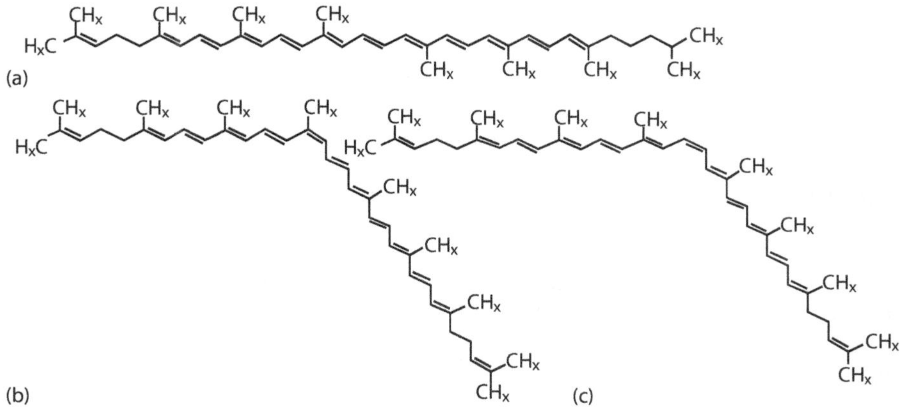

FIGURE 9.3 Chemical structure of lycopene all-trans (a) and cis-isomers (b and c).

of 40 carbon atoms (tetra terpenes), having precisely numerous conjugated double bonds responsible for the characteristic color of these pigments, red and orange, respectively, due to the absorption of light at a wavelength of 400–500 nm able to excite electrons delocalized along the polyene chromophore [42]. A carotenoid, due to the presence of numerous double bonds in the molecule, can show up in a wide variety of geometrical isomers in particular in configurations all-trans (E), cis (Z) mono isomers, or polyisomers. The predominant form in fresh tomatoes is all-trans lycopene but trans-cis isomerization is formed during processing and storage operations (Figure 9.3). The all-trans lycopene differs from Z isomers for the molecular structure, stability, and light absorption properties. Its molecule is completely flat and relatively stable [43].

9.8 VITAMIN C

Vitamin C is an antioxidant molecule present in many fruit and vegetables. It is water-soluble and quite unstable in aqueous solution. Among the vitamins contained in tomatoes, vitamin C or ascorbic acid (Figure 9.4) is that present in much higher amounts, at concentrations that generally range, in the cultivars used for industrial processing, from 30 to 50 mg/100 g of fresh fruit [44]. As happens for lycopene, vitamin C also needs intake through food because the body does not synthesize it. Ascorbic acid, being a reducing agent with a powerful antioxidant effect, acts

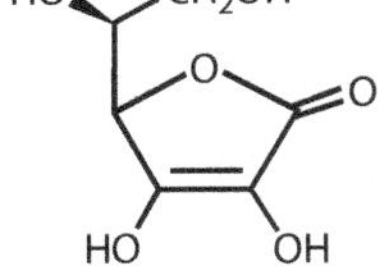

FIGURE 9.4 Structure of ascorbic acid (AA).

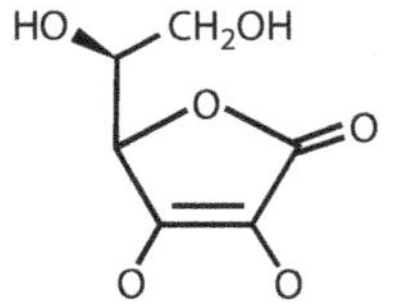

FIGURE 9.5 Structure of dehydroascorbic acid (DHAA).

as radical scavenger, reducing and consequently neutralizing oxygen reactive species (hydrogen peroxide) or counteracting the effects of toxic substances. As it is one of the least stable vitamins, losses can be experienced during processing, cooking, or storing. It is quite stable under acidic conditions, typical of the processing of tomato products, in its oxidized form: L-dehydroascorbic acid (dehydroascorbic acid, DHAA) (Figure 9.5) [45].

9.9 POLYPHENOLS

Polyphenols (oligo elements of the bioflavonoid type), are potent natural antioxidants present in plants that, because of their chemical structure, are useful in preventing oxidation of lipoproteins and in "sequestering" free radicals. Experimental data show that polyphenols lead to a strengthening of endogenous antioxidant defences and that this is achieved through the activation of antioxidant responsive elements, ARE, involved in the induction of antioxidant and de-toxicant enzymes [46].

Mainly phenolic compounds are divided into phenolic acids and flavonoids. The latter can be divided into several subclasses such as flavonols, flavones, flavanones, flavanols, anthocyanidins, and isoflavones [47].

Flavonoids (Figure 9.6) are pigments widespread in vegetables, deriving biosynthetically from phenylalanine. They have two aromatic rings, which include a heterocyclic ring; this structure coincides with that of the Flavin molecule, which, according to the alkylation, glycosylation, and hydroxylation reactions it may undergo, gives rise to the seven subclasses of flavonoids previously mentioned. The advantage of the chemical structure of flavonoids is to have a high antioxidant activity in both hydrophilic and lipophilic systems. There are studies showing that flavonoids in combination with vitamin E are able to protect LDLs (low-density lipoproteins) from oxidation.

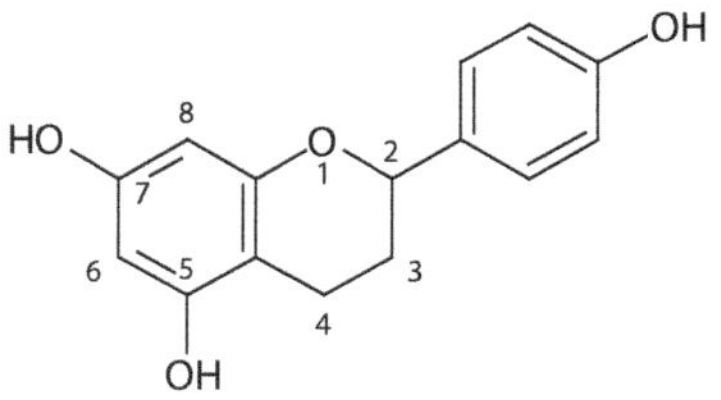

FIGURE 9.6 General structure of flavonoids.

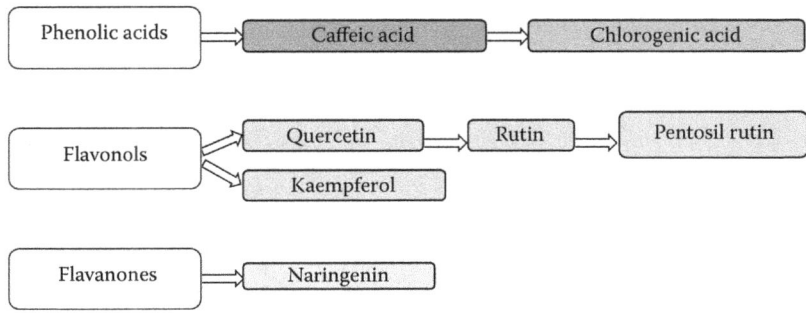

FIGURE 9.7   Principal polyphenols of tomato products.

Recent studies have shown that the polyphenol content in processing tomato (Figure 9.7) undergoes only a slight decrease after various heat treatments, which makes all products from industrial processing particularly interesting as to the total final content in primary antioxidant compounds (lycopene, polyphenols) [48]. Polyphenols are a family of about 8,000 organic molecules and, in tomato, their formation (naringin, quercetin, coumaric acid, etc.) under normal conditions is favored by high solar radiation [49]. Along with terpenes, they are the most widespread antioxidants in the category of non-nutrients; among nutrients, there are vitamins C and E.

Their antioxidant power is correlated to the current number of phenolic rings, and to the number and position of hydroxyl groups and double bonds present in the molecule. In particular, their antioxidant power is determined by the presence of a de-hydroxylated B-ring (catechol group), by unsaturation in the 2.3 position associated with a 4-carbonyl function in the C-ring or by functional groups able to chelate transition metals. Moreover, structural differences lead to differences in the bioavailability of these compounds, both in terms of different absorption in the gastrointestinal tract and of different metabolism and distribution capacity in tissues and organs [50] (Figures 9.7.1 and 9.7.2).

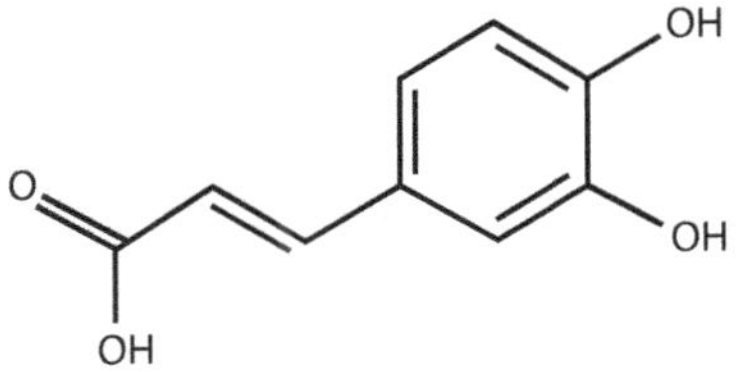

9.7.1.1 Structure of caffeic acid 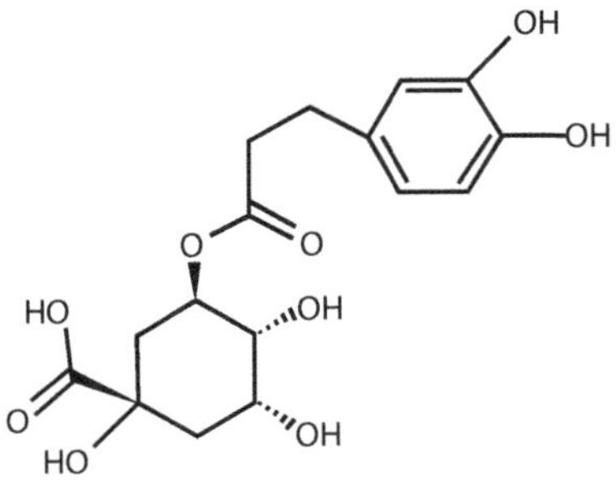   **9.7.1.2 Structure of chlorogenic acid**

FIGURE 9.7.1   Structure of phenolic acids.

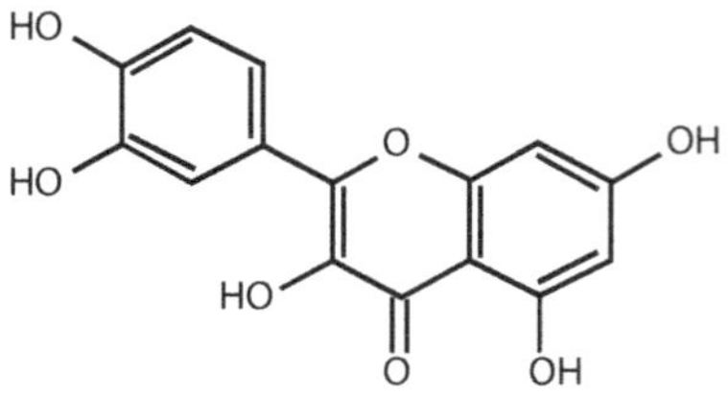

9.7.2.1 Structure of quercetin

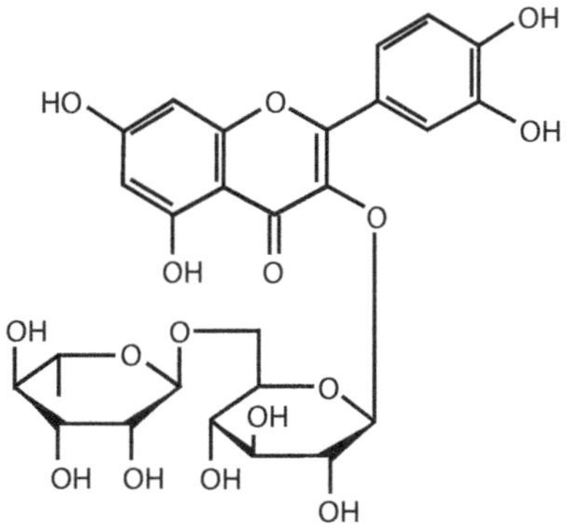

9.7.2.2 Structure of rutin

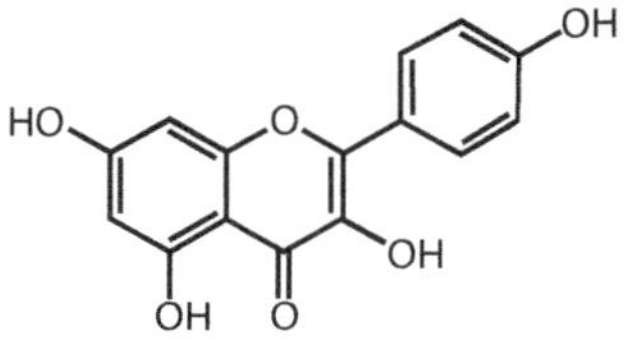

9.7.2.3 Structure of kaemferol

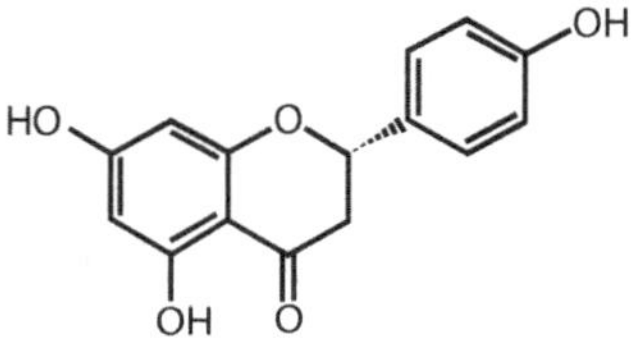

9.7.2.4 Structure of naringenin

FIGURE 9.7.2   Structure of tomato flavonols and flavanones.

9.10 STABILITY OF BIOACTIVE TOMATO MOLECULES DURING FOOD PROCESSING

The results of many studies and the evidences quoted in the media have pushed towards the application of new research in order to verify the effective concentration of the bioactive molecules and their bioavailability behavior along all the tomato production chain: from fresh matter until finished products (Figures 9.8 and 9.9) [51–53]. Lycopene content in tomatoes changes on the basis not only of genetic and cultural factors but also of ripening stage. Varietal variability for processing tomatoes (red and ripe) ranges between 2,000 and 3,500 mg/kg dry matter [54]. During the different kinds of industrial processing, the decrease in lycopene content is very low (less than 8%), except for drying processes; in this case a decrease was found of about 10% in the case of powders obtained by spray drying and/or in dry tomatoes [55].

Lycopene is not destroyed during food processing and heat improves bioavailability. Consumption of tomato products with oil improves biovailability [56].

It has been clearly demonstrated that lycopene absorption and bioavailability by the human organism is higher if the tomatoes ingested have been previously treated by physical thermal or enzymatic actions resulting in cell wall destruction. Indeed,

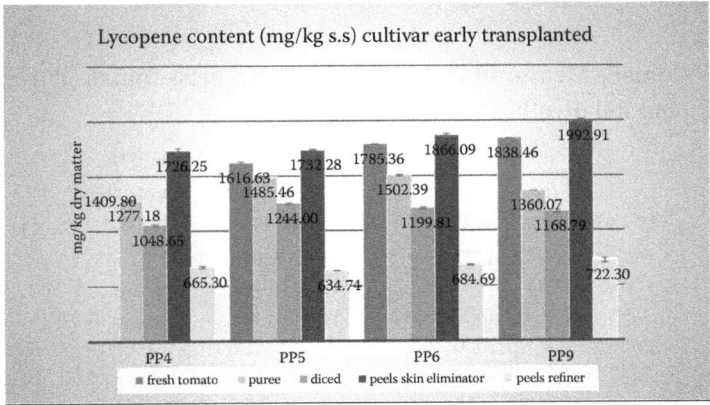

FIGURE 9.8   Lycopene content of early transplanted cultivar (mg/kg dry matter).

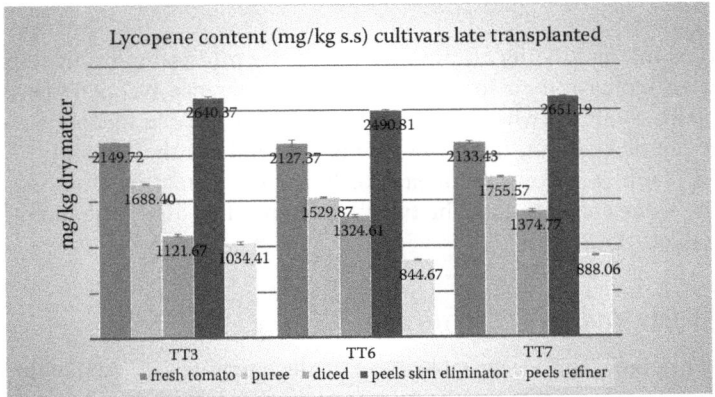

FIGURE 9.9   Lycopene content of late transplanted cultivar (mg/kg dry matter).

absorption is greater for heat-treated tomato products as compared with fresh tomatoes or untreated tomato products, in which cell walls have been destroyed (cold-break single-strength and concentrated juices in comparison with hot-break single-strength and concentrated juices). The physical state and the kind of techno-logical treatment, which causes a certain degree of matrix destruction, play a very important role in lycopene reactivity and in its bioavailability [57].

9.10.1   Lycopene Behaviour during Processing

As already reported in previous papers [24], the distribution of lycopene content in the two types of products (diced tomatoes and tomato purée) was found to be consistently greater in the purée. That behavior was observed for products obtained from both early-transplanted (PP) and late-transplanted (TT) cultivars (Table 9.1). The presence of seeds, fibrous parts, yellow shoulders, and discoloured

TABLE 9.1

Lycopene Content Measured in Fresh Tomato Samples, Tomato Purée, Diced Tomatoes, and Relative Byproducts

	Lycopene Content (mg/kg dry matter)				
Cultivars	Fresh Tomato	Purée	Diced	Peels Skin Eliminator	Peels Refiner
PP4	1409,8 ± 11,9	1277,2 ± 7,3	1048,6 ± 10,2	1726,2 ± 25,9	665,30 ± 9,1
PP5	1616,6 ± 10,8	1485,5 ± 8,2	1244,0 ± 8,0	1732,3 ± 7,8	634,74 ± 0,5
PP6	1785,4 ± 2,7	1502,4 ± 6,8	1199,8 ± 8,9	1866,1 ± 14,5	684,69 ± 5,9
PP9	1838,5 ± 2,4	1360,1 ± 0,7	1168,8 ± 12,6	1992,9 ± 19,5	722,30 ± 21,4
TT3	2149,7 ± 0,7	1688,4 ± 7,9	1121,7 ± 12,7	2640,4 ± 16,0	1034,41 ± 15,9
TT6	2127,4 ± 36,4	1529,9 ± 8,4	1324,6 ± 13,9	2490,8 ± 9,6	844,67 ± 4,4
TT7	2133,4 ± 14,9	1755,6 ± 7,1	1374,8 ± 14,8	2651,2 ± 2,9	888,06 ± 8,1
MEANS ± DS	**1865,8** ± 273,6	**1514,1** ± 107,9	**1211,7** ± 107,9	**2157,1** ± 400,2	**782,02** ± 138,5

woody cores that, not being eliminated during the processing of diced tomatoes, contribute to increase their total solids content, decrease lycopene percent concentration. To establish whether the difference between the means is significant (that is, it can be said that the difference is not random) the Student "t" test was performed. Table 9.1 shows the means of lycopene content by coupling for each cultivar the values obtained for the two types of product analyzed and of the relative byproducts obtained.

9.10.2 TOTAL VITAMIN C CONTENT

Vitamin C (ascorbic acid) is unstable under acidic conditions, typical of the processing of tomato products. Among the many vitamins present in tomatoes, vitamin C is that present in much higher quantities. In the cultivars used for industrial purposes, its concentration generally ranges from 20 to 30 mg/100 g of fresh fruit (400–600 mg/100 g dry matter). Vitamin C is principally concentrated close to the mesocarp, in the portion near the pericarp of the fruit; this is probably due to a precise influence by the light, like the other important bioactive substances of the tomato (polyphenols and carotenoids). From the results obtained it has been noticed that the major vitamin C decrease seems to be due to the technological processing treatments conducted at high temperatures. Vitamin C is sensitive to heat in the presence of oxygen and its degradation is catalyzed by the presence of metals such as copper. In the modern equipment used by the canning industry, almost entirely made of stainless steel and operating under conditions controlled enough to avoid contact with oxygen, this inconvenience is greatly reduced; nevertheless, too high a shelf life temperature and other processing operations with oxygen inclusion could cause a reduction in vitamin C.

Table 9.2 and Figures 9.10 and 9.11 show the results relating to total vitamin C content.

TABLE 9.2
Vitamin C Content: Fresh Tomatoes, Tomato Purée, Diced Tomatoes, and Relative Byproducts

	Vitamin C Content (mg/100 g dry matter)				
Cultivars	Fresh Tomato	Purée	Diced	Peels Skin Eliminator	Peels Refiner
PP4	524,62 ± 2,63	283,32 ± 2,86	271,28 ± 0,94	119,76 ± 1,56	54,34 ± 0,53
PP5	509,27 ± 4,74	282,83 ± 1,80	274,36 ± 0,53	123,88 ± 2,19	69,53 ± 3,08
PP6	520,90 ± 12,24	287,76 ± 2,17	283,43 ± 1,55	122,17 ± 4,39	68,56 ± 1,43
PP9	499,43 ± 9,84	265,78 ± 0,33	257,23 ± 1,76	113,09 ± 2,65	66,79 ± 0,15
TT3	623,57 ± 5,02	333,77 ± 0,37	293,35 ± 2,85	108,02 ± 0,25	64,99 ± 3,38
TT6	599,62 ± 13,85	319,31 ± 2,54	290,12 ± 2,07	113,87 ± 1,65	63,70 ± 0,05
TT7	612,58 ± 11,09	324,08 ± 1,25	276,41 ± 1,53	114,70 ± 0,45	66,49 ± 0,40
MEANS ± DS	**555,71 ± 51,50**	**299,55 ± 24,50**	**278,03 ± 11,74**	**116,50 ± 5,68**	**64,91 ± 5,04**

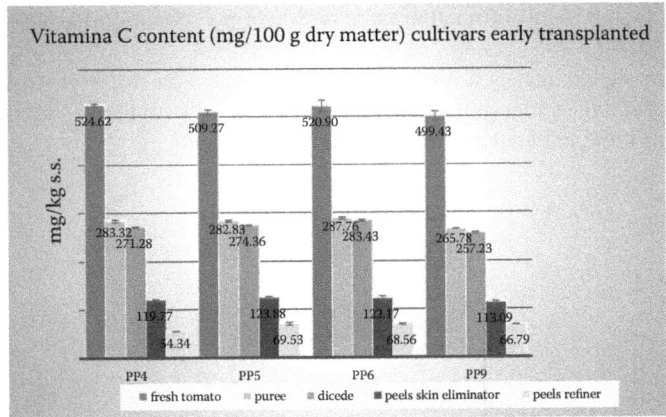

FIGURE 9.10 Vitamin C content: early transplanted cultivar (mg/kg dry matter).

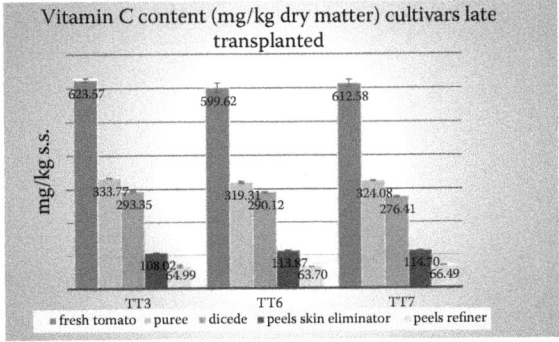

FIGURE 9.11 Vitamin C content: late transplanted cultivars (mg/kg dry matter).

9.10.3 TOMATO PHENOLIC CONTENT

The bibliography shows that the most important polyphenols present into tomatoes are six characteristic molecules: naringenin, quercetin, rutin, kaempherol, caffeic acid, and chlorogenic acid.

For chromatographic assessments, a linear gradient HPLC/DAD method is usually applied. In order to validate the method a repeatability test was carried out (five extractions were performed on the same sample). Repeatability is expressed as variation coefficient (standard deviation/mean ×100). A calibration curve was obtained by measuring the absorbance of different levels of concentration of the standard solutions (1, 5, 10, 20, and 50 ppm) with optimal linearity ($R^2 > 0.99$). The chromatographic peaks were integrated at characteristic max absorbance wavelength for each standard. Repeatability for rutine standard was found to be $0.02/10.45 = 0.19\%$.

The recovery trial was peformed (in quintuplicate) by adding to the same sample an aliquot of the solution at a known concentration of rutine (50 ppm). On the sample with standard addition were performed both the extraction according to the modalities of the above procedure and the determination of the concentration of principal polyphenols by HPLC. The recovery percentage of rutine standard was 97.77.

Figure 9.12 shows the chromatographic profile of polyphenols analyzed into refiner peels byproducts.

Tomato purée maintains a total phenolic concentration (referred to 1 kg of dry matter) comparable to that of the fresh matter. This is due to the partial incidence of skins that are not completely removed in the refiner step.

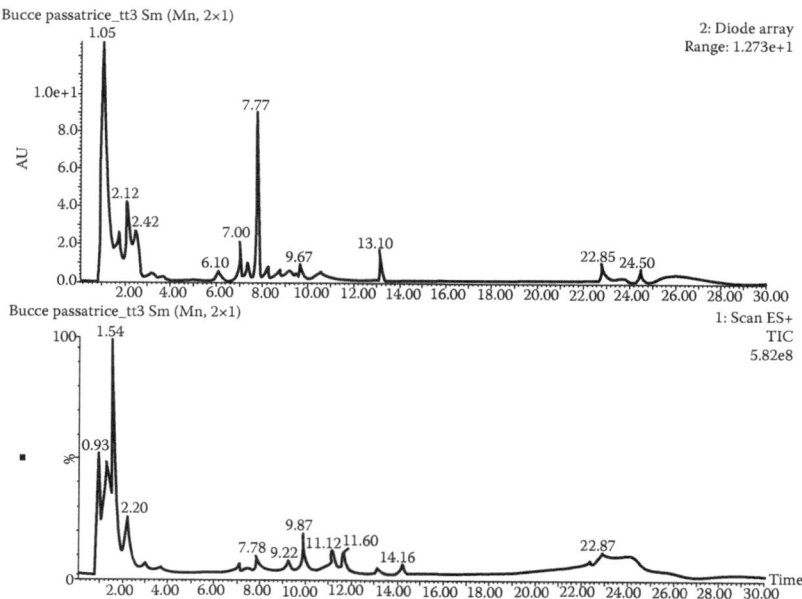

FIGURE 9.12 Polyphenolic chromatographic profile of refiner peels. Above: chromatogram of UV scansion (280–440 nm). Below: chromatogram of MS scansion (100–800 Da).

TABLE 9.3

Phenol Content Measured in Fresh Matter, Tomato Purée, and Diced Tomatoes of Early-Transplanted Cultivars

Cultivars Early-Transplanted (mg/kg dry matter)	Fresh				Purée				Diced			
	Pentosyl-Rutin	Rutin	Naringenin	Quercetin	Pentosyl-Rutin	Rutin	Naringenin	Quercetin	Pentosyl-Rutin	Rutin	Naringenin	Quercetin
PP4	204,35	532,63	6,95	3,15	151,14	532,83	132,63	<loq	49,28	181,73	30,49	<loq
PP5	260,82	663,44	11,21	3,16	211,91	650,71	117,64	<loq	88,65	279,15	55,56	<loq
PP6	277,88	865,59	15,41	6,33	226,55	885,25	230,44	<loq	81,54	259,58	55,52	<loq
PP9	214,38	504,77	6,70	<loq	218,06	601,58	95,93	<loq	65,74	227,60	47,01	<loq
MEANS	239,36	641,61	10,07	4,21	201,92	667,59	144,16	<loq	71,30	237,02	47,14	<loq
dev. std.	32,98	152,42	3,87	1,70	31,83	141,63	55,07		17,47	39,39	11,50	<loq

TABLE 9.4
Phenol Content Measured in Solid Byproducts of Early-Transplanted Cultivars

Cultivar Early-Transplanted (mg/kg dry matter)	Peels From Skin Eliminator				Peels From Refiner			
	Pentosyl-Rutin	Rutin	Naringenin	Quercetin	Pentosyl-Rutin	Rutin	Naringenin	Quercetin
PP4	423,84	1397,86	85,26	<loq	103,80	431,95	24,02	<loq
PP5	645,79	1699,02	102,28	11,98	102,73	352,05	19,62	1,81
PP6	763,98	241,84	138,90	233,45	111,87	438,72	51,82	8,38
PP9	721,95	1596,99	92,43	23,06	131,40	416,78	23,66	2,59
MEANS	638,89	1233,93	104,72	89,49	112,45	409,87	29,78	4,26
dev.std.	140,31	623,20	22,22	111,70	12,46	37,50	13,77	3,21

TABLE 9.5

Phenol Content Measured in Fresh Matter, Tomato Purée, and Diced Tomatoes of Late-Transplanted Cultivars

Cultivar Late-Transplanted (mg/kg dry matter)	Fresh				Purée				Diced			
	Pentosyl-Rutin	Rutin	Naringenin	Quercetin	Pentosyl-Rutin	Rutin	Naringenin	Quercetin	Pentosyl-Rutin	Rutin	Naringenin	Quercetin
TT3	91,00	349,83	11,11	<loq	101,89	347,15	121,31	<loq	48,20	165,63	28,78	<loq
TT6	198,93	478,10	10,40	<loq	204,09	519,03	69,14	<loq	76,04	213,21	21,59	<loq
TT7	288,02	603,78	7,03	<loq	119,88	251,98	53,38	<loq	73,53	171,85	28,63	<loq
MEANS	192,65	477,24	9,51	<loq	141,95	372,72	81,28	<loq	65,92	183,56	26,33	<loq
dev. std.	88,27	113,63	1,96	<loq	48,80	121,07	31,80	<loq	13,90	23,58	4,06	<loq

TABLE 9.6

Phenol Content Measured in Solid Byproducts of Late-Transplanted Cultivars

Cultivar Late-Transplanted (mg/kg dry matter)	Peels From Skin Eliminator				Peels From Refiner			
	Pentosyl-Rutin	Rutin	Naringenin	Quercetin	Pentosyl-Rutin	Rutin	Naringenin	Quercetin
TT3	217,45	678,68	90,50	4,61	66,20	283,67	28,69	1,16
TT6	573,30	1550,70	59,21	<loq	178,21	569,16	18,39	1,88
TT7	340,90	318,43	51,01	<loq	97,67	283,10	12,91	<loq
MEANS	377,22	965,60	66,91	4,61	114,03	378,64	19,99	1,52
dev. std.	161,62	453,29	18,77	1,55	51,67	147,69	7,39	0,41

Tables 9.3 through 9.6 show the single phenolic concentration measured into fresh tomatoes, tomato purees, diced tomatoes, and relative tomato byproducts produced.

Since tomato purée and diced tomatoes showed an undoubtedly different dry matter content, all analytical results obtained were referred to 1 kg of dry matter to obtain a more significant appraisal.

Figure 9.13 shows the average content of major tomato polyphenols and their single performance evaluated during processing.

Finally, to better discriminate higher antioxidant content samples, and therefore with higher nutritional interest, all the data were set up in a graphic (Figure 9.14) whose histograms bars were divided in three sections representative each the single contribute of the antioxidant compounds: lycopene (darkest grey), vitamin C (grey), and total phenols (dark grey). Statistical analysis reveals that there are no significant differences between the different analyzed samples.

A decrease in lycopene and in total phenolic compounds assessed in processed products is mainly due to the partial elimination of the peels during the technological operations to obtain final product recipes (purée and peeled diced tomatoes). On the contrary, vitamin C, that is, the thermal sensitive compound, is merely highly influenced by the processing temperature and by the oxygen present during the processing operations. Thanks to these indications, it has been possible to reveal the extremely important amounts of bioactive/antioxidants compounds that are in fact discharged.

The composite co-presence of the key antioxidants lipophilic characteristics substances like lycopene (and all the other tomato carotenoids), together with the partial hydro soluble substances, like Vitamin C and phenolic compounds (flavonoids, flavones and hydrossicinnamic conjugated acids), have to be considered the responsible of the preventive and protective effects of the tomato products upon the consumer's health. Nevertheless, it remains clear that further evidence and investigations are required to better study the biochemical behavior of these molecules and understand the link between cellular and metabolic effect of tomato components and the

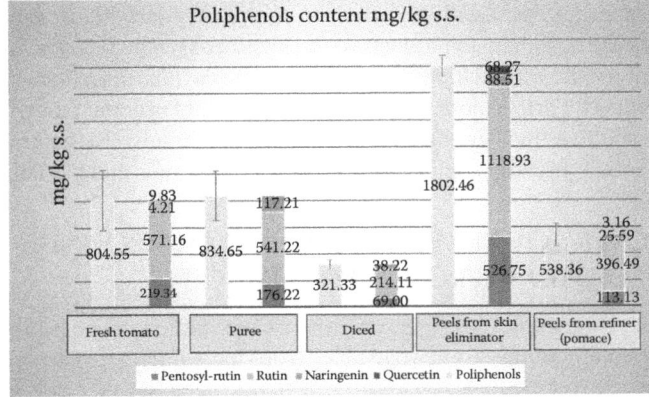

FIGURE 9.13 Tomato phenolic (average content) measured in fresh tomatoes, purée, diced tomatoes and byproduct (mg/kg dry matter).

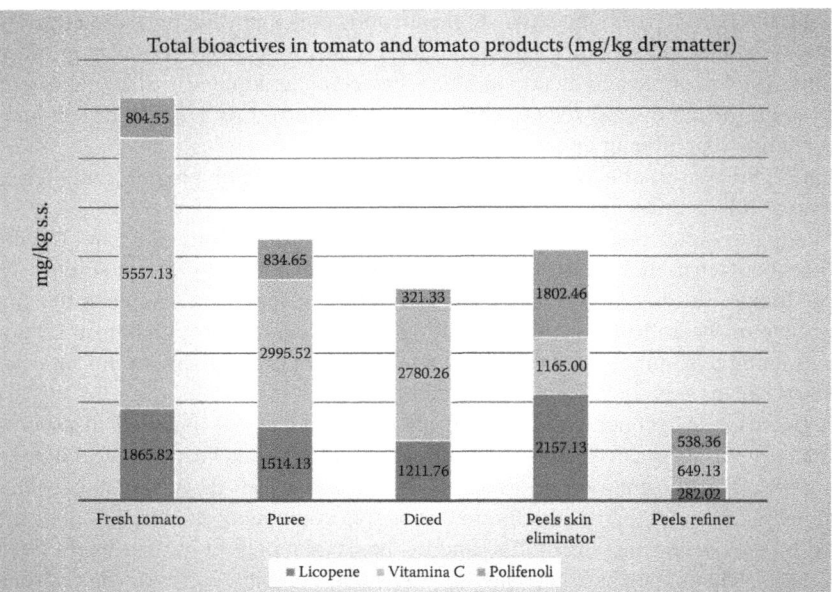

FIGURE 9.14 Total bioactive tomato compounds: concentration of lycopene, vitamin C, and phenols (mg/kg dry matter) assessed in fresh matter, tomato purée, diced tomatoes, and related byproducts.

bioavailability for the tomato products trying to confirm and finally successfully demonstrate the synergic effect in biological human action.

In SSICA, since the year 2000, some research protocols have been planned to assess the "tomato bioactive antioxidants performance" to appraise some procedures for tomato byproduct recycling. This could be of interest both from an environmental viewpoint and for the health benefits of all bioactive tomato compounds. The processing tomato industry could optimize its productions by means of adding natural components in order to obtain products with higher fiber consistency and higher color intensity. Moreover, processing tomato byproducts (peels and seeds) could be converted into very cheap material for the formulation of new nutraceuticals, pharmaceuticals, and finally functional products. Furthermore, treated byproducts could become "supernatural" colorants for the additives industry and a potential source of innovative natural substances for environmentally friendly applications (packaging materials and bio-compost fuels).

9.11 RECOVERY OF BIOACTIVE TOMATO COMPOUNDS FROM SOLID BYPRODUCTS (POMACE)

During recent years, numerous research groups have studied the possibility of recovering bioactive compounds from vegetable byproducts and/or from their industrial processes. Recently a SSICA research project has been set in order to introduce

TABLE 9.7
Lycopene Content in Selected High-pigment Tomato Cultivars

Cultivar	Brix	Lycopene mg/kg Fresh Matter	Lycopene mg/kg Dry Matter
CLX38196	4.85	148	3051.55
DEXTER	5.36	161	3003.73
ISI 2561	5.32	166	3120.30
NPT66	6.03	183	3034.83
H9478	5.45	169	3100.92
LITTANO	5.47	165	3016.45
NPT111	5.53	165	2983.73
TRAJAN F1	4.92	155	3150.41
UGX822	4.95	155	3131.31
WALLYRED	5.02	165	3286.85

an innovative process for the production of new tomato-based juice (*"beta-juice"*) extract, naturally enriched in bioactive components (lycopene and polyphenols) improved from selected high-pigment (round-square-shape) tomato cultivars (Table 9.7). Data concerning this enriched tomato beta-juice extract obtained that it could be easily employed both as source of natural antioxidant colorant useful for the formulation of novel functional food and/or as a *concentrated bioactive* additive to be used for functional food fortification. Moreover, this experimental process has proved to be in agreement with the emerging concepts of sustainability and recycle/reuse requirements. This technology involves a complete "Eco-sustainable" thermophysical and mechanical extraction treatment that leads to concentrate and enhances the bioactive compounds that are typically found in processed products (e.g., pulp and purée).

The first part of the work has been elected at the selection of suitable processing tomato varieties rich in bioactive compounds, followed by the application of a proper setup of a novel pilot plant able to extract and concentrate (only by thermophysical bases) the natural bioactive compound coming from juice and tomato pomace mixed and well homogenized together (Beta-Juice*). The research has provided a specific analytical evaluation of possible pesticide residues and qualitative industrial parameters (°Brix, dry matter, total acidity, pH, sugar content, hunter color) and nutraceutical substances (lycopene, polyphenols, ascorbic acid). Data analyses are shown in Tables 9.8 and 9.9. Moreover, studies were made testing the application, during tomato pulp and purée production, by the addition of 10% and 15% of homogenized "beta-juice" to both products and their quality and compositional characterization. Results are shown in Table 9.10.

TABLE 9.8

Qualitative and Nutritional Composition of "Tomato Beta-Juice*"

Beta Juice*	Total Solids %	Soluble Solids %	pH	Total Acidity %	Total Sugars %	Total Sugars/ T.S. %	Ascorbic Acid mg/ kg/T.S.	Lycopene mg/kg/T.S.	Total Phenols mg/kg/T.S.
After extraction	5.28	4.33	4.35	0.38	2.49	47.16	1423.00	4688.00	1221.90
After concentration	9.01	5.60	4.28	0.47	2.55	28.30	0.00	9657.00	1202.72

*Juice and tomato pomace extract.

TABLE 9.9
Pesticide Residues of "Tomato Beta-Juice*"

Pesticides	Analytical Technique	Unity of Measure Quantification Limit (mg/kg)	Beta-Juice Sample
a) Fungicides	LC/MS/MS	0,005	< L.Q.
b) Fungicides	GC/MS/MS	0.005	< L.Q.
c) Erbicides	GC/MS/MS	0.005	< L.Q.
d) Piretroides	GC/MS/MS	0.005	< L.Q.
e) Naturals	LC/MS/MS	0.010	< L.Q.

*Juice and tomato pomace extract.

Data concerning the experimental enriched tomato purée (fortified tomato products) evidently shows that this experimental novel technology could be employed as a natural source of antioxidants for the formulation of novel functional foods with contemporary results of high sustainability and recycle/reuse (circular economy). In particular, the addition of 10% and 15% of beta-juice* in a common commercial purée, allowed a significant improvement respectively of the 60% and 70% of total lycopene in the final product. Food industry raw material, surplus productions, byproducts, and solid wastes and wastewaters are usually mostly discharged, and this behavior significantly reduces the increasing demand of sustainability for the food industry. The same unused food industrial byproducts would become, after proper treatment with natural biological or "green" techniques, very cheap fonts of new fine chemicals (such as antioxidants, vitamins, etc.) and/or novel natural food bioactive additives (e.g., fibers, enzymes, natural pigments, etc.).

The deep development in mild technologies and the introduction of new environmentally friendly techniques and novel applications enabled the extraction, production, separation, and concentration of natural bioactive compounds that could be easily used in the nutraceutical – cosmetic – food additive markets. The above experimental technology example involves a complete "eco-sustainable" extraction process that leads to obtaining bulk products with high bioactive compound concentration without any solvent treatment.

Tomatoes and, above all, their industrial byproducts will become a very cheap raw material alternative for the production of natural bio-active compounds for the food and non-food markets in the future. Further studies will be carried out in order to industrialize this new "green technology" to the entire vegetable processing industry in order to enhance the momentum of the circular economy.

Recent advances in structural and functional genomics, as well as advances in plant breeding, biotechnology applications, and the introduction of the best available green processing, enable the creation of new high value-added functional foods specifically tailored for the health and the wellbeing of modern consumers.

TABLE 9.10

Qualitative and Nutritional Composition of "Fortified Products"

	Total Solids %	pH	Total Acidity %	Total Acidity/ T.S. %	Total Sugars %	Total Sugars/ T.S. %	Lycopene mg/kg	Lycopene mg/kg/T.S.	Color			
									L	a	b	a/b
Commercial	7.15	4.27	0.39	5.45	3.22	44.9	119	1668	23.67	27.96	13.73	2.04
Tomato Purée	7.18	4.28	0.38	5.29	3.22	41.4	116	1618	23.49	28.04	13.65	2.05
No addition	7.16	4.24	0.42	5.87	3.18	44.4	112	1568	23.44	27.98	13.68	2.04
Average	**7.16**	**4.26**	**0.40**	**5.54**	**3.21**	**43.6**	**115**	**1618**	**23.53**	**27.99**	**13.69**	**2.04**
Commercial	7.76	4.33	0.41	5.28	3.40	43.3	200	2547	23.52	29.09	13.71	2.12
Tomato Purée +	7.82	4.31	0.42	5.37	3.36	42.8	210	2675	23.54	29.10	13.71	2.12
10% Beta-juice*	7.97	4.34	0.41	5.14	3.40	43.3	200	2547	23.49	29.09	13.69	2.13
Average	**7.85**	**4.33**	**0.41**	**5.27**	**3.39**	**43.1**	**203**	**2590**	**23.52**	**29.09**	**13.70**	**2.12**
Commercial	7.78	4.33	0.43	5.53	3.30	42.4	213	2746	23.52	29.22	13.67	2.14
Tomato Puree +	7.76	4.31	0.40	5.15	3.32	42.7	215	2765	23.54	29.28	13.66	2.14
15% Beta-juice*	7.78	4.33	0.41	5.27	3.30	42.4	213	2746	23.50	29.25	13.67	2.14
Average	**7.77**	**4.32**	**0.41**	**5.32**	**3.31**	**42.5**	**214**	**2753**	**23.52**	**29.25**	**13.67**	**2.14**

*Juice and tomato pomace extract.

REFERENCES

1. Hybertson B.M., Gao B., Bose S.K., McCord J.M. Oxidative stress in health and disease: The therapeutic potential of Nrf2 activation. *Mol Aspects Med* 32 (2011):234–46.
2. Ahuja A., Singh N., Gupta P., Mishra S., Rani V. Influence of exogenous factors on skin aging. In: Farage M.A., Miller K.W., Maibach H.I. (eds). *Textbook of Aging Skin*, ISBN 978-3-662-47397-9. Springer-Verlag, Heidelberg (2016).
3. Corpet D.E., Gerber M. Alimentation méditérranéenne et Santé. I-caractéristiques. Maladies cardio-vasculaires et autres affections. *Méd Nut* 4 (1997):129–42.
4. Oroian M., Escriche I. Antioxidants: Characterization, natural sources, extraction and analysis. *Food Res Int* 74 (2015):10–36.
5. Liu R.H. Health benefits of fruit and vegetables are from additive and synergistic combinations of phytochemicals. *Am J Clin Nutr* 78(suppl) (2003):517S–20S.
6. Ferrari Carlos K.B. Functional foods and physical activities in health promotion of aging people. *Maturitas* 58(4) (2007):327–39.
7. Giovannini C., Filesi C., D'Archivio M., Scazzocchio B., Santangelo C., Masella R. Polifenoli e difese antiossidanti endogene: Effetti sul glutatione e sugli enzimi ad esso correlati. *Ann Ist Super Sanità* 42(3) (2006):336–47.
8. Carratù B., Sanzini E. Sostanze biologicamente attive presenti negli alimenti di origine vegetale. *Ann Ist Super Sanità* 41(1) (2005):7–16.
9. Allen Stevens M. Relationships between components contributing to quality variation. *J Am Soc Hort Sci* 97 (1972):70.
10. Georg S., Brat P., Alter P., Amiot M.J. Rapid determination of Polyphenols and Vitamin C in Plant- Derived Products. *J Agric Food Chem* 53 (2005):1370–3.
11. Nicoli M.C., Calligaris S., Marzocco L. Effect of enzymatic and chemical oxidation on the antioxidant capacity of catechin model system of apple derivatives. *J Agric Food Chem* 48 (2000): 4576–80.
12. Hermann K. Zur quantitativen Veraenderung phenolisher inhaltsstoffe bei der gewinnung vin Apfel—und Birnensaeften. *Fluessiges Obst* (1993): 7–10.
13. Viana M., Barbas C., Bonet B., Bonet M.V., Castro M., Fraile M.V., Herrera E. In vitro effects of a flavonoid-rich extract on LDL-oxidation. *Atherosclerosis* 123 (1996): 82–91.
14. Gould W.A. *Tomato Production, Processing & Technology* (3rd ed). *CTI Publications Inc* (1992).
15. *Tomato News.* WPTC: Crop update as of 2 November 2016. No. 12/2016, 28th Year. ISSN 1145-9565 Tomatoland Infor. Service, AMITOM. December 2016.
16. Sandei L., Leoni C. Exploitation of by-products (solid wastes) from tomato processing to obtain high value antioxidants. *ISHS Acta Horticulturae* 724: IX International Symposium on the Processing Tomato (2006).
17. Burton-Freeman B.M., Sesso H.D. Whole food versus supplement: Comparing the clinical evidence of tomato intake and lycopene supplementation on cardiovascular risk factors. *Am Soc Nutr Adv Nutr* 5 (2014): 457–85.
18. Rout S.P., Choudary K.A., Kar D.M., Das L., Jain A. Plants in traditional medicinal system-future source of new drugs. *Int J Pharm Pharm Sci* 1(1) (July–Sep. 2009).
19. Vulcano D., Sandei L., Leoni C. Contributo specifico del licopene all'attività antiossidante della frazione liposolubile dei derivati del pomodoro. *Industria Conserve* 77 (2002):219–39.
20. Capanoglu E., Beekwilder J., Boyacioglu D., Hall R., de Vos R. Changes in antioxidant and metabolite profiles during production of tomato paste. *J Agric Food Chem* 56(3) (2008):964–73.
21. Navarro-González I., Pérez-Sánchez H., Martín-Pozuelo G., García-Alonso J., Periago M.J. The inhibitory effects of bioactive compounds of tomato juice binding to hepatic HMGCR. In vivo study and molecular modelling. *PLoS ONE* 9(1) (2014).

22. Vallverdú-Queralt A., Jáuregui O., Medina-Remón A., Lamuela-Raventós R.M. Evaluation of a method to characterize the phenolic profile of organic and conventional tomatoes. *J Agric Food Chem* 60(13) (2012):3373.
23. Shi J. (ed.). *Functional Food Ingredients and Nutraceuticals Processing Technologies* (2nd ed.). CRC Press, Boca Raton, FL (2015) pp. 609–38.
24. El-Agamey A., Lowe G.M., McGarvey D.J., Mortensen A., Truscott G., Young A.J. Carotenoid radical chemistry and antioxidant/pro-oxidant properties. *Arch Biochem Biophys* 430(1) (2004): 37-48.
25. Barrett D.M., Weakley C., Diaz J.V., Watnik M. Qualitative and nutritional differences in processing tomatoes grown under commercial organic and conventional production system. *J Food Chem Toxicol* 12 (2007):441–452.
26. Giovannucci E. Tomatoes, tomato-based products, lycopene, and cancer: Review of the epidemiologic literature. *J Natl Cancer Inst* (1999): 317–31.
27. Agarwal S., Rao A. Tomato lycopene and its role in human health and chronic diseases. *Can Med Assoc J* 163 (2000):739–44.
28. Rao A.V., Ray M.R., Rao L.G. Lycopene. *Adv Food Nutr Res* 51 (2006):99–164.
29. Schierle J., Bretzel W., Bühler I., Faccin N., Hess D., Steiner K., Schüep W. Content and isomeric ratio of lycopene in food and human blood plasma. *Food Chem* 59 (1997):459–65.
30. Sandei L., Siviero P., Zanotti G., Cabassi A., Leoni C. Evaluation of the Lycopene content in processing tomato cultivars claiming "high pigment content." ISHS. *Acta Hort* (613) (2003):331–3.
31. Lenucci M.S., Caccioppola A., Durante M., Serrone L., De Caroli M., Piro G., Dalessandro G. Carotenoid content during tomato (*Solanum lycopersicum* l.) fruit ripening in traditional and high-pigment cultivars. *Ital J Food Sci* 21(4) (2009).
32. Leoni C. I derivati industriali del pomodoro Collana di monografie tecnologiche Stazione Sperimentale per l'Industria delle Conserve Alimentari in Parma 7 (1993).
33. Kilambi H.V., Kumar R., Sharma R., Sreelakshmi Y. Chromoplast-specific carotenoid-associated protein appears to be important for enhanced accumulation of carotenoids in hp1 tomato fruits. 161 (2013). American Society of Plant Biologists.
34. Clinton S.K., Emenhiser C., Schwartz S.J., Bostwick D.G., Williams A.W., Moore B.J., Erdman J.W. Jr., cis-Trans lycopene isomers, carotenoids, and retinol in the human prostate. *Cancer Epidemiol Biomark Prev* 5 (1996):823–33.
35. Stahl W., Junghans B., de Boer E. S., Driomina E. S., Briviba K., Sies H. Carotenoid mixtures protect multilamellar liposomes against oxidative damage: Synergistic effects of lycopene and lutein. *FEBS Lett* 427 (1998):305–8.
36. Agarwal S., Rao A.V. Tomato lycopene and low density lipoprotein oxidation: A human dietary intervention study. *Lipids* 33 (1998):981–4.
37. Stahl W., Sies H. Uptake of lycopene and its geometrical isomers is greater from heat-processed than from unprocessed tomato juice in humans. *J Nutr* 122(11) (1992):2161–6.
38. Pérez-Conesa D., García-Alonso J., García-Valverde V., Iniesta M.D., Jacob K., Sánchez-Siles L.M., Ros G., Periago M.J. Changes in bioactive compounds and antioxidant activity during homogenization and thermal processing of tomato puree. *Innov Food Sci Emerg Technol* 10 (2009):179–88.
39. Abushita A., Daood H.G., Biacs P.A. Change in carotenoids and antioxidant vitamins in tomato as a function of varietal and technological factors. *J Agric Food Chem* (2000).
40. Ferreira A.L.A., Corrêa C.R. Lycopene bioavailability and its effects on health. In: Lima G., Vianello F. (eds.). *Food Quality, Safety and Technology*. Springer, Vienna (2013).
41. Grolier P. Composition of tomatoes and tomato products in antioxidants. *The White Book on the Antioxidants in Tomatoes and Tomato Products and Their Health Benefits*, 2000 Final Report CT 97-3233 CMITI Sarl—Tomato News Supplement.

42. Dewick P.M. *Chimica, biosintesi e bioattività delle sostanze organiche naturali* (1997).
43. Bramley P.M. Strategies for alteration of carotenoid levels in tomato. *Proceedings of the Tomato & Health Seminar. Pamplona* (1998), pp. 95–100.
44. Giuntini D., Graziani G., Lercari B., Fogliano V., Soldatini G.F., Ranieri A. Changes in carotenoid and ascorbic acid contents in fruits of different tomato genotypes related to the depletion of UV-B radiation. *J Agric Food Chem* 53 (2005):3174–81.
45. Di Matteo A., Sacco A., Anacleria M., Pezzotti M., Delledonne M., Ferrarini A., Frusciante L., Barone A. The ascorbic acid content of tomato fruits is associated with the expression of genes involved in pectin degradation. *BMC Plant Biol* 10 (2010):163.
46. Crozier A., Lean M.E.J., McDonald M.S., Black C. Quantitative analysis of the flavonoid content of commercial tomatoes, onions, lettuce and celery. *J Agric Food Chem* 45 (1997):590–5.
47. Sandei L., Vadalà R. Accelerated solvent extraction applied to tomato polyphenols assessment. *ACTA Hort* 971 (2013):225–33.
48. Raffo A., La Malfa G., Fogliano V., Maiani G., Qualglia G. Seasonal variations in antioxidant components of cherry tomatoes (*Lycopersicon esculentum* cv. Naomi F1). *J Food Compos Anal* 19(1) (2006):11–9.
49. Rice-Evans C.A., Miller N.J., Pagania G. Structure-antioxidant activity relationships of flavonoids and phenolic acids. *Free Rad Biol Med* 20 (1996):933–56.
50. Chanforana C., Loonisa M., Moraa N., Caris-Veyrat C., Dufoura C. The impact of industrial processing on health-beneficial tomato microconstituent. *Food Chem* (2012).
51. Capanoglu E., Beekwilder J., Boyacioglu D., De Vos R.C.H., Hall R.D. The effect of industrial food processing on potentially health-beneficial tomato antioxidants. *Crit Rev Food Sci Nutr* 50 (2010):919–30.
52. Vallverdú-Queralt A., Medina-Remón A., Andres-Lacueva C., Lamuela-Raventos R.M. Changes in phenolic profile and antioxidant activity during production of diced tomatoes. *Food Chem* 126(4) (2011):1700–7.
53. Fuentes E., Forero-Doria O., Carrasco G., Maricán A., Santos L.S., Alarcón M., Palomo I. Effect of tomato industrial processing on phenolic profile and antiplatelet activity. *Molecules* (2013).
54. Sandei L., Siviero P., Zanotti G., Cabassi A., Leoni C. Evaluation of the lycopene content in proc. tomato cultivars claiming "high pigment content." *Acta Hort* (613) (2003).
55. Leoni C., Sect. WG2—Effects of mechanical and thermal treatments and storage conditions on antioxidant content and their bioavailability in processed tomatoes. *The White Book on the Antioxidants in Tomatoes and Tomato Products and Their Health Benefits*, 2000 Final Report CT 97-3233 CMITI Sarl—Tomato News Supplement.
56. Fielding J.M., Rowley K.G., Cooper P., O'Dea K. Increases in plasma lycopene concentration after consumption of tomatoes cooked with olive oil. *Asia Pac J Clin Nutr* 14(2) (2005):131–6.
57. Sandei L., Risi P., Mezzadri V., Vietta V. Evaluation of the content in bioactive antioxidants of industrial tomato products: tomato puree and diced tomatoes. *Ind Cons* 85 (2010):91–109.

10 Lycopene
Antioxidant Health Claims and Regulation

*Montaña Cámara, Virginia Fernández-Ruiz,
Laura Domínguez, Rosa María Cámara,
and M. Cortes Sánchez-Mata*

CONTENTS

10.1 INTRODUCTION

In the near future, food will not only be needed to meet the needs of population growth, but these foods must be more nutritious and serve to alleviate nutritional deficiencies. This is one of the objectives of the World Health Organization (WHO), since health is considered "a state of complete physical, mental and social well-being, and not merely the absence of disease or disease" (WHO 1990).

Bioactive compounds may have very varied properties, structures, and functions (Halliwell 1987; Lampe 1999). Among the various bioactive compounds present in foods of plant origin, which contribute to the prevention of diseases and to the improvement of the quality of life of the population, are carotenoids such as lycopene (abundant in tomatoes, watermelons and pink varieties of grapefruit), which has anti-oxidant activity and is involved in reducing the risk of suffering from different types of cancer and cardiovascular disease.

According to the ILSI (1999), a Functional Food is "one that contains a component, nutrient or non-nutrient, with selective activity related to one or more functions of the organism, with an added physiological effect above its nutritional value and whose positive actions justify that their functional (physiological) or even healthy character may be claimed."

In Europe, functional foods have to fulfill the following characteristics: being able to be consumed daily or as part of the normal diet; being constituted by natural ingredients, in some cases in a concentration higher than that found in the original food or containing some ingredient absent in the original food; the beneficial effects have to be scientifically proven and must go beyond those derived from their nutritional value; they should improve the condition of the individual, improve the quality of life or reduce the risk of suffering any pathology. The presence of these bioactive compounds in significant quantities in foods is what justifies their qualification as "functional foods," although this is not an official qualification, since there is no legislation that supports it.

10.2 EUROPEAN REGULATION ON HEALTH CLAIMS

The European Union established on 19 January 2006 the European Regulation on nutrition and health claims in labeling (Regulation (EC) 1924/2006), which prohibits a food from being promoted as having therapeutic or curative properties, and establishes the following categories of statements: "nutritional statements" or "content," "health claims," and "disease risk reduction statements." Health claims are expressions that describe a relationship between a food substance and a disease or other health related condition (e.g., a "risk reduction" relationship). They are defined as any voluntary message or commercial representation in any form such as text, declaration, image, logo, etc., that affirms, suggests or implies that there is a relationship between the food object of the claim and health and the type of claim subject to evaluation.

The requirement of this regulation is that any statement should be based on scientific proven and real evidence, and applied to nutrition and health claims made in commercial communications; whether in the labeling, presentation, or advertising of foods supplied as such to the final consumer. In essence, the Regulation ensures the protection of the right of consumers to truthful, proven information with a rigorous scientific basis—a particularly relevant aspect in the case of food.

The statements that refer to health are as follows:

- Declarations of reduction of disease risk (Article 14). They are those that imply or suggest that the consumption of a food significantly reduces a risk factor for the appearance of a human disease whose relationship with food intake has been properly documented.
- Health claims other than those for the reduction of disease risk or health claims (Article 13). They describe functions in the body and do not refer to diseases or pathological conditions.
- Declarations concerning the development and health of children (Article 14).

The general conditions to be met by an allegation of this type is that its beneficial effect is demonstrated by scientific evidence, that the nutrient or substance is in the final product in significant quantities to produce the claimed effect that is bioavailable, and that the quantity of product that could reasonably be consumed in a balanced diet provides a significant amount of the declared nutrient or substance. In addition, these statements should be understandable to the average consumer and will refer to ready-to-eat foods.

The application process is carried out through EFSA and, in particular, the scientific evaluation is the responsibility of the NDA Panel (Panel on Dietetic Products, Nutrition and Allergies) within the Nutrition Unit. In the evaluation of proposals submitted for the acceptance of health claims, the NDA Panel considers that the food or product in question is well defined and characterized, that the claimed effect is clear, definite, and has a beneficial physiological effect in terms of human health. It requires that the cause-effect relationship between the consumption of the food or constituent and the alleged effect be well established (for the target group or population and under the proposed conditions of use). Moreover, the wording of the proposed claim must clearly reflect what has been scientifically proven, meet the criteria set out in the Regulation, and that the recommended amounts of the product or food required obtaining the claimed effect can be consumed within a balanced diet. The allegations refer to healthy people, and therefore the effects that are expected are much more limited than in cases of illness, because it means "improving or maintaining the health of people considered healthy."

The health claims submitted that have been accepted are listed in the Annex to Regulation 432/2012, as are the list of health claims that may be attributed to food (other than those relating to the reduction of disease risk and the development and health of children) and its subsequent amendments.

The following definitions shall apply: "CLAIM" means any message or representation, which is not mandatory under Community or national legislation, including pictorial, graphic, or symbolic representation, in any form, which states, suggests or implies that a food has particular characteristics; "NUTRIENT" means protein, carbohydrate, fat, fibre, sodium, vitamins, and minerals listed in the Annex to Directive 90/496/EEC, and substances that belong to or are components of one of those categories; "OTHER SUBSTANCE" means a substance other than a nutrient that has a nutritional or physiological effect.

This Regulation specifies that the term "nutrient" includes proteins, carbohydrates, fats, fibers, sodium, vitamins, and minerals listed in the Annex to Directive 90/496/EEC and Directive 2008/100/EC, as well as substances belonging to one of these categories or components of one of them. "Other substance" refers a substance other than a nutrient, which has a nutritional or physiological effect.

With these considerations in relation to lycopene, since it is not a nutrient, it would fall under the category of "other substances" and there was no explicit authorization of its possible use in labeling under the category of "health claims." Until now, the only permissible claim in relation to lycopene contained in its products is "naturally contains lycopene" or its variations.

Types of health claim:

- The so-called "Function Health Claims" (or Article 13 claims): relating to the growth, development, and functions of the body; referring to psychological and behavioural functions; on slimming or weight-control.
- The so-called "Risk Reduction Claims" (or Article 14(1)(a) claims) on reducing a risk factor in the development of a disease. For example: "Plant stanol esters have been shown to reduce blood cholesterol. Blood cholesterol is a risk factor in the development of coronary heart disease."
- Health claims referring to "children's development" (or Article 14(1)(b) claims). For example: "Vitamin D is needed for the normal growth and development of bones in children."

The requirement of the Regulation is that any statement must be based on proven scientific evidence. It applies to nutrition and health claims made in commercial communications, whether in the labeling, presentation, or advertising of foods to be delivered as such to the final consumer.

This work is focused on reviewing the up-to-date EFSA currently approved antioxidant health claims related to lycopene, as well as the rejected applications.

10.3   EFSA REQUIREMENTS TO JUSTIFY THE ANTIOXIDANT HEALTH CLAIMS

According to EFSA recommendations it would be necessary to carry out appropriate studies and/or measures to justify the proposed claims with the following considerations: conduct human intervention studies with healthy individuals; studies should be well controlled (specifying control/placebo well); use the markers considered as valid by EFSA quantifying according to the appropriate analytical methodology for this agency; or use new, properly documented markers (Cámara et al. 2012b). The following considerations should be also taken into account:

Risk factor in relation to incidence of the disease. The identification of a risk factor for claims to reduce disease risk is a requirement in Regulation (EC) 1924/2006. Data on reducing the incidence of disease can be used as evidence of the allegations function if the disease shows a clear dysfunction of a particular organ or tissue (e.g., data on cases of coronary disorders can be used to justify a health claim concerning the maintenance of normal heart function).

Justified effects versus long-term effects. In general, measurements of the variable to be considered at different time points during the intervention study are required. Short-term studies (e.g., three or four weeks) can also be considered for the scientific basis of these claims.

The reproducibility of the effect. Multi-center studies can be used to illustrate the reproducibility of the effect, provided that each center has sufficient capacity to evaluate it. However, multicenter trials are generally designed in order to achieve a sufficient sample size to show a significant effect on the study variable.

Antioxidant defence. The induction of antioxidant enzymes cannot in itself be used as evidence of claims related to the "antioxidant defence system" for constituents of non-nutrients, such as lycopene.

Oxidative damage. The autoantibodies circulating against LDL are not direct markers of oxidative lipid damage but an indirect measure of immune system function, so its association with oxidative lipid damage is not justified (Miller et al. 2005). The spectrophotometric method TBARS and MDA (modification of the TBARS method including HPLC) are frequently used to determine lipid oxidative damage, as indicated in the related references PASSCLAIM (Mensink et al. 2003). While frequently used, these markers have not been validated in the context of the scientific substantiation of health claims. As such, they are not direct methods of measuring lipid oxidative damage *in vivo*. The TBARS spectrophotometric assay is not specific and its lack of specificity has rendered it obsolete. The HPLC modification of the TBARS method separating the MDA-TBA adduct from interfering compounds has improved sensitivity, specificity, and reproducibility (Griffiths et al. 2002; Knasmüller et al. 2008), and can be used as supporting evidence for the scientific substantiation of health claims relating to lipid protection from oxidative damage (in addition to measures of F2-isoprostanes and oxidation of LDL *in vivo*). Similarly, lipid oxidation products (peroxides or hydroperoxides), as well as derivatives of phospholipids (phosphatidylcholine-hydroperoxides), can be measured by HPLC and used (as MDA analyzed by HPLC) as supporting evidence for health claims on the protection of oxidative damage lipids.

Food–drug interaction. Claims on the synergistic or additive effects of foods when a medicinal product is consumed are not evaluated under Regulation (EC) No 1924/2006.

Adequate control groups for human studies on blood lipids. It is necessary to include an appropriate control group in accordance with the study design to take into account the confounding factors that may affect the relationship between food/constituent consumption and the reported effect.

Assessment of blood pressure (systolic versus diastolic). The isolated decrease in systolic blood pressure is always beneficial, while a reduction in diastolic blood pressure may be beneficial if it is accompanied by a reduction in systolic blood pressure.

Flow-mediated dilation. The flow-mediated dilation (FMD) technique is an example of a well-established technique in *in vivo* assays to evaluate endothelium-dependent vasodilation.

Metabolism of homocysteine. The Panel considers maintaining normal homocysteine levels as a positive physiological effect.

10.4   ANTIOXIDANT PROPERTIES OF LYCOPENE

Oxidative DNA damage is considered to be a causative factor for various types of cancer, and fruit and vegetables, due to their high antioxidant content, can be considered important chemo preventive agents.

Antioxidant compounds prevent the negative effects of free radicals on tissues and fats, reducing the risk of cancer and cardiac abnormalities by avoiding the *in vitro* oxidation and cytotoxicity of molecules such as LDL, thereby decreasing atherogenicity (Kong et al. 2010; Seelert 1992).

Many of the bioactive compounds, such as lycopene, have antioxidant capacity, and are therefore able to counteract the oxidative stress caused by the attack of highly oxidizing molecules, such as free radicals, to different tissues and biomolecules of the organism (genetic material, plasma, and membrane lipoproteins), causing cellular aging processes, and the appearance of cardiovascular diseases, cancer, cataracts, or neurological disorders, among others.

Free radicals are produced as a consequence of the aerobic activity of the cells, are highly reactive, and can damage a large number of biological molecules. Between 1% and 3% of the oxygen consumed by the body's cells is transformed into ROS (reactive oxygen species), which are potentially harmful. An important function of antioxidants is to intercept triplet states, to prevent the formation of singlet oxygen, or to trap it directly. They are also reactive with other oxygen species such as hydroxyl radical and superoxide anion. In the organism, there are defense mechanisms against oxidation such as the enzymes glutathione peroxidase and reductase, catalase, or superoxide dismutase.

When the balance between free radicals and antioxidants is lost in favor of the former, harmful processes are triggered that are associated with the development of numerous diseases such as cardiovascular alteration, initiation of cancerous processes, formation of cataracts, aging processes, inflammatory processes, and neurological disorders. One of the most common diseases related to the cardiovascular system is atherosclerosis, which is characterized mainly by the progressive obstruction of the arteries as a consequence of the accumulation of lipids in the arterial wall. These lipids cross the endothelium, accumulating and oxidizing in endothelial cells, vascular smooth muscle cells, and macrophages. It has been reported that oxidized LDL, and also its degradation products, may be involved in atherosclerosis in both initiation and progression of the disease. In addition, other processes related to the development of atherosclerosis, such as inflammation, cell proliferation, and thrombosis, may also have their origin in oxidation or be induced by oxidized lipoproteins (Berliner and Heinecke 1996).

One of the most important families of antioxidants present in the diet are the carotenoid (as the case of lycopene). Substances included in the denomination of carotenoids can be only synthesized in the plants, and only arrive at animal tissues through feeding process; after that carotenoids can be modified or accumulated.

Lycopene is the main carotenoid in tomato and responsible of its characteristic red color. Epidemiological results correlate the ingestion of this carotenoid with a reduction of different types of cancer, such as digestive system or prostate cancer, as well as a reduction of coronary diseases—one of the main causes of mortality in developed countries.

The presence of a system of nine or more conjugated connections confers to lycopene its capacity to absorb light in the UV-visible region. Its chemical structure captures reactive oxygen species (ROS) working as an antioxidant at low oxygen pressure. This antioxidant action is beneficial in the prevention and improvement of certain pathologies, since most of them start due to a cellular oxidation process.

All these properties are documented in the book *Tomatoes, Lycopene and Human Health* (Rao 2016), a good compendium of all aspects of the healthy properties of lycopene in tomato and tomato products, their mechanisms of action, and impact on health.

10.5   LYCOPENE HEALTH CLAIMS STATUS

After examining more than 44,000 requests for claims submitted by individual Member States, EFSA published on its website in May 2010 the consolidated list of claims relating to the "general function and health." Of these, EFSA has published more than 300 opinions and provided scientific advice on more than 2,500, of which 33 referred to lycopene as either a specific compound, component of a food, or constituent of a mixture in a commercial product. One corresponds to Article 14 and most of them correspond to Article 13, of which 25 correspond to Article 13.1 requests, related to the antioxidant properties of lycopene. There are also four claims on botanical substances for which finalization is pending.

Within lycopene, the only authorized claim is the following: "contains natural lycopene" or variations always providing the term "naturally" or "natural" as a prefix to the statement.

Although, there is abundant literature related to the antioxidant properties of lycopene. In relation to health claims, so far the only allowed claim was for tomato derivatives corresponding to a tomato concentrate without lycopene, for which the claimed effect is the "reduction of platelet aggregation," corresponding to Article 13(5), EFSA (2009a). Other claims requested as: Article 13(1) Lycopene antioxidant properties (EFSA 2011) and Article 14: Reduction of risk disease (EFSA 2009b) were rejected.

The list of permitted health claims established by Regulation (EU) No 432/2012 is regularly updated with newly authorized health claims (http://ec.europa.eu/food /safety/labelling_nutrition/claims/health_claims/index_en.htm).

In the following section the 33 applications related to lycopene and corresponding to articles 13.1 (25 applications), 13.5 (three applications), and 14 (one application) are discussed. There are also four claims on botanical substances for which finalization is pending.

10.5.1 LYCOPENE HEATH CLAIMS APPLICATIONS—ARTICLE 13.1

With regard to the requests under Article 13(1), as shown in Table 10.1, the applications were mostly focused on:

- Prevention of oxidative damage to DNA, proteins, and lipids.
- Protecting the skin against oxidative damage.
- Maintaining cardiac function.
- Maintaining vision.

In March 2011, EFSA issued a statement in relation to all proposals relating to the antioxidant properties of lycopene included in Article 13(1), denying them in the terms in which they were drafted (EFSA 2011a).

TABLE 10.1
Lycopene Health Claims Applications Under Article 13(1)

ID Code	Health Claim Proposed
	Lycopene Prevention of Oxidative Damage to DNA, Proteins and Lipids
ID 1608:	Lycopene and antioxidant properties.
ID 1611:	Lycopenes from tomato juices and antioxidant properties.
ID 1899:	Lycopenes from tomato pulp and sauces and antioxidant properties.
ID 1942:	Lycopenes from tomato juice and oxidative stress control.
ID 1663:	Tomato extract containing lycopene and antioxidant properties/cell and DNA protection.
ID 2081:	Lycopene (from tomato extract) and antioxidant properties.
ID 2082:	Lycopene (from tomato extract) and antioxidant properties/protection of DNA. BASF.
ID 2142:	Standardized tomato extract for antioxidant protection system/protection of DNA.
ID 1662:	Tomato extract containing lycopene and maintenance of cardiovascular health
ID 1609:	Lycopene and prostate health.
ID 1664:	Tomato extract containing lycopene and maintenance of prostate health.
ID 2374:	Tomato extract containing lycopene and prostate health.
	Lycopene Skin Protection Against Oxidative Damage
ID 1259:	Guava and skin health.
ID 1607:	Lycopene and skin health. Up to 16 mg/d. 405-Ferrosan.
ID 1665:	Tomato extract containing lycopene and maintenance of skin health.
ID 2143:	Standardized tomato extract for skin health.
ID 2262:	Guava and skin health.
ID 2373:	Tomato extract containing lycopene and skin health.
	Lycopene Maintenance of Cardiac Function
ID 1610:	Lycopene and heart health. (409, DSM) 40–60 mg.
ID 2372:	Tomato extract containing lycopene and cardiovascular health.
	Lycopene Maintenance of Vision
ID 1827:	Lycopene and eyes.

The claimed effects were "antioxidant properties," "antioxidant properties/ cellular protection and DNA," "antioxidant properties/DNA protection," "oxidative stress control," "antioxidant protection/Maintains cardiovascular health," "prostate health," and "maintenance of prostate health." In the absence of any specification, the target population is understood to be the general population. In relation to the claim of "antioxidant properties," the panel assumes that the alleged health effects relate to the protection of DNA, proteins, and lipids from oxidative damage. Reactive oxygen species (ROS), including various types of radicals, are generated in biochemical processes (e.g., respiratory chain) and, as a result of exposure to exogenous factors (e.g., radiation and pollutants). These reactive intermediates can cause damage to molecules such as DNA, proteins, and lipids, if not intercepted by the antioxidant network, which includes free radical scavengers. This means that the antioxidant properties of lycopene only produce a beneficial physiological effect by providing protection against the oxidative damage of DNA, proteins, or lipids.

EFSA considered that tomatoes and their tomato derivatives are the main dietary sources of lycopene, along with other foods and synthetic lycopene, recently authorized in the EU as a new food ingredient. In addition, EFSA considered that lycopene is sufficiently characterized and that protection of DNA, proteins, and lipids from oxidative damage may be a beneficial physiological effect. However, it considered that none of the reviewed studies fully complies with the requirements of EFSA to support the claims of antioxidant properties of lycopene, antioxidant capacity, and antioxidant defence system. The panel based the justification for the rejection of all applications submitted in the consideration that none of the studies provided information regarding the significant effect of lycopene consumption on reliable markers of oxidative damage to DNA, lipids, or proteins as compared to a control. EFSA believes that human studies are central to the justification of health claims. Regarding the types of intervention studies needed to justify the effect, the population should consist of healthy individuals, and studies should be well controlled (specifying control/placebo well).

Some experiments, and generally small-scale human studies, investigate the effects of lycopene consumption on the total antioxidant activity of plasma measured by various assays such as: the measurement of antioxidant capacity according to Trolox equivalents (TEAC), the total antioxidant reactive potential (TRAP), the chemiluminescence assay using 2,2′-azinobis (3-ethylbenzothiazoline-6-sulfonate) (ABTS) as a reagent. Autoantibodies circulating against LDL are not direct markers of oxidative lipid damage but an indirect measure of immune system function, thus its association with oxidative lipid damage is not justified (EFSA 2011a).

The spectrophotometric methods TBARS and MDA (modification of the TBARS method including HPLC) are frequently used to determine lipid oxidative damage, although, in spite of their frequent use, these markers have not been validated in the context of the scientific basis of statements of healthy properties. As such, they are not direct methods of measuring lipid oxidative damage *in vivo*. The TBARS spectrophotometric assay is not specific and its lack of specificity has rendered it obsolete.

The HPLC modification of the TBARS method separating the MDA-TBA adduct from interfering compounds has improved sensitivity, specificity, and reproducibility, and can be used as supporting evidence for the scientific substantiation of health claims relating to the protection of lipids from oxidative damage (in addition to F2-isoprostane measurements and LDL oxidation *in vivo*). Similarly, lipid oxidation products (peroxides or hydro peroxides), as well as derivatives of phospholipids (phosphatidylcholine-hydroperoxides), can be measured by HPLC and used (as MDA analyzed by HPLC) as supporting evidence for the bedding scientific claims of health on the protection of oxidative damage lipids.

The panel considered that evaluation of total antioxidant activity/plasma potential, and/or GSH concentration, and/or antioxidant enzymatic activities are not markers of oxidative damage; that the formation of MDA measured as TBARS, as well as the resistance of LDL to oxidation, are not suitable markers for assessing lipid peroxidation; that variants of the comet assay to measure DNA damage do not reflect oxidative DNA damage but independent breaks in the DNA strands; that the *ex vivo* resistance of DNA to oxidation does not reflect oxidative DNA damage *in vivo*; and that measurement of thiol protein concentrations is not a reliable marker of oxidative damage to proteins. In addition, several of the human studies presented corresponded to uncontrolled intervention trials with lycopene or tomato derivatives (EFSA 2011a).

The panel further considered that the claims on the antioxidant capacity of lycopene in food were based on its ability to eliminate free radicals evaluated *in vitro* on model systems and that the information provided does not indicate that lycopene exerts an effect beneficial physiology in humans, as required by Regulation 1924/2006. Therefore, in the weighting of the evidence, the panel found that none of the studies provided provide information regarding the significant effect of lycopene consumption on reliable markers of oxidative damage to DNA, lipids, or proteins as compared to a control. Conclusions could not be drawn from these references for the scientific basis of the declared effect.

Rao and Agarwal (1998) conducted a randomized, crossover study with five different sources of lycopene in 19 healthy subjects (nine women and ten men), non-smokers aged between 25 and 40 years. The duration of the study was one week, with each source of lycopene providing 20.5 mg/d (spaghetti sauce A), 39.2 mg/d (spaghetti sauce B), 50.4 mg/d (tomato juice), 75 mg/d (tomato A oleoresin) and 150 mg/d (tomato B oleoresin). Controls received a placebo from which no data were given. As a marker of DNA damage, 8-oxodG was measured on lymphocytes by HPLC with CE detection. For lipid peroxidation, TBARS determination was performed in serum, and protein oxidation was estimated by the loss of free serum thiol groups measured with beta dystrobrevin (5,5'-dithio-bis (2-nitrobenzoic) beta, DTNB). No statistically significant differences were found in DNA oxidation between the different groups and placebo. The panel found that TBARS and the loss of serum free thiol groups measured with DTNB are not reliable markers of oxidative damage on lipids and proteins, respectively.

TABLE 10.2
Other Applications of Lycopene Under Article 13(1)

ID 1202:   Tomato juice and cardiovascular system.

ID 1796:   Carotenoids (alpha, beta, and gamma carotene, lycopene) and skin.

ID 1859:   Soy isoflavones + lycopene + zinc + selenium + vitamin D + vitamin E + vitamin C and
             sexual organs and/or hormone activity.

TABLE 10.3
Claims on Botanical Substances for Which Finalization Is Pending

ID 1666:   Tomato extract, grape seeds extract, vitamins C and E, Selenium (Seresis Pharmaton) and
             antioxidant combination, for antioxidant protection system.

ID 1667:   Tomato extract, grape seeds extract, vitamins C and E, Selenium (Seresis Pharmaton) and
             cardiovascular health.

ID 1668:   Tomato extract, grape seeds extract, vitamins C and E, Selenium (Seresis Pharmaton) and
             skin anti-ageing agent.

ID 1669:   Combination of lycopene, proanthocyanidins, vitamin C, vitamin E, selenium, and
             beta-carotene and contribution to normal collagen formation and protection of the skin
             from UV-induced damage. Seresis Pharmaton.

Other lycopene health claims were submitted according to Article 13(1), and all of them were rejected by EFSA (see Table 10.2); four claims were classified as botanical substances for which finalization is pending (Table 10.3).

10.5.2   Lycopene Heath Claims Applications—Article 13.5

With regard to the lycopene applications under Article 13(5), those based on newly developed scientific evidence and/or for which protection of proprietary data is requested, the following two applications were rejected.

- Lycopene, vitamin E, lutein, and selenium and protection of the skin from UV-induced damage. Applicant: Competent Authority of Cyprus, following an application by Nutrilinks Sarl. Rejected, September 27, 2012, December 14, 2012.
- Combination of blackcurrant seed oil, fish oil, lycopene, vitamin C, and vitamin E, and helps to improve dry skin conditions. Applicant: Laboratoires Innéov SNC. Rejected, May 21, 2010.

In May 2009, the application submitted by Provexis Natural Products and corresponding to "Water-soluble tomato concentrate (WSTC I and II) and platelet aggregation" was approved. The food component subject to the Provexis health claim is a lycopene-free, fat-free, water-soluble tomato concentrate developed in two forms: WSTC I, a fully water-soluble syrup, and its low-sugar derivative WSTC II, in powder form. The alleged effect is "reduced platelet aggregation," where the target population is composed of healthy adults between 35 and 70 years of age. The panel considered that a cause-and-effect relationship had been established between the consumption of water-soluble tomato concentrate and the reduction of platelet aggregation in humans. On the other hand, to achieve the stated effect, 3 grams of WSTC I or 150 milligrams of WSTC II should be consumed daily in up to 250 ml of fruit juices, flavoured drinks, or yoghurt drinks (EFSA 2009a).

10.5.3 Lycopene Heath Claims Applications—Article 14

Regarding the requests of Article 14 (reduction of disease risk factor), only the claim "Prevent oxidative damage of plasma lipoproteins has been presented, which reduces the accumulation of plaques in the arteries and reduces the risk of disease heart disease, stroke and other clinical complications of atherosclerosis" was proposed. It was rejected in August 2009 (EFSA 2009b).

Regardless of the specific allegations of lycopene mentioned above, it should be taken in mind that since December 14, 2012, foodstuffs marketed in the European Union, including tomato derivatives, must comply with the following rules in relation to claims in labeling: R. 1924/2006; and their subsequent amendments. To date, health claims whose use is permitted are those contained in Commission Regulation 432/2012.

10.6 EFSA-VALID MARKERS FOR OXIDATIVE DAMAGE

10.6.1 Markers of Oxidative Damage to Lipids

Direct evaluation of oxidative lipid damage (lipid peroxidation) can be obtained *in vivo* by measuring changes in F2-isoprostanes in 24-hour urine samples, which is a better matrix than plasma for this measurement, and using gas chromatography, or preferably, mass spectrometry. F2-isoprostanes can also be measured using specific immunoassays. However, lack of specificity due to possible cross-reactions with other prostanoids needs to be taken into account.

Measurements of lipid oxidative damage (e.g., lipid peroxidation) could also be obtained *in vivo* by measuring the oxidized LDL particles in the blood using immunological methods (e.g., antibodies) with appropriate specificity. Phosphatidylcholine hydroperoxides (PCOOH) measured in blood or tissue by HPLC are also an acceptable markers.

Other proposed methods are not reliable for markers of lipid peroxidation *in vivo* (e.g., thiobarbituric acid reactive substances (TBARS), malondialdehyde (MDA), lipid peroxides associated with HDL-paraoxonases, conjugated dienes, hydrocarbons, autoantibodies against LDL), and *ex vivo* (oxidation resistance of LDL). However, concentrations of MDA or lipid peroxides in blood or tissue could be used as supportive evidence (e.g., in addition to measurements of F2-isoprostanes and *in vivo* oxidation of LDL) if appropriate techniques are used for analysis (e.g., HPLC).

10.6.2 MARKERS OF OXIDATIVE DAMAGE TO PROTEINS

Direct measurements of oxidative damage to proteins *in vivo* (e.g., measurement of oxidative changes of amino acids in proteins) can be obtained by means of HPLC-MS techniques and other methods, as long as the identification and separation of such molecules in the plasma from other substances is successfully resolved (e.g., tyrosine nitration protein products).

Measurements of oxidation byproducts of proteins (e.g., protein carbonyls) using standard assays (e.g., colorimetric procedure involving dinitrophenylhydrazine (DNPH) derivatization of carbonyl groups) or ELISA methods (either directly or after derivatization DNPH) are generally susceptible to interference by other molecules, and may only be used in combination with at least one direct marker of oxidative protein damage *in vivo* if evaluated directly in blood or target tissue (e.g., skin).

10.6.3 MARKERS OF OXIDATIVE DAMAGE TO DNA

Direct evaluation of oxidative DNA damage can be obtained *in vivo* using modifications of the comet assay, which allow the detection of oxidized DNA bases (e.g., the use of endonuclease III to detect oxidized pyrimidines). Although the assay does not provide absolute values, it allows quantitative comparison with an appropriate control. This assay directly reflects oxidative DNA damage within cells, for example, in circulating lymphocytes. Measurement of DNA damage using the traditional kite assay (single-cell microgel electrophoresis, SCGE) detecting DNA strand breaks by the appearance of the tail is not specific for oxidative damage. Other variants of the comet assay determining resistance against *ex vivo* oxidative modification are also not suitable methods for assessing oxidative DNA damage *in vivo*.

Quantification of 8-hydroxy-2-deoxyguanosine (8-OHdG) in blood (e.g., lymphocytes), tissue (e.g., skin) and urine have been used to evaluate oxidative DNA damage. 8-OHdG is a result of oxidative damage and repair by cleavage, but may also result from oxidation of free bases or nucleotides, oxidation of other nucleic acids, and artifacts during the preparation of the sample. The urine presence of 8-OHdG does not directly reflect DNA oxidation within cells, but could be used in combination with direct measurement of oxidative DNA damage if appropriate techniques (HPLC) are used.

10.6.4 STUDIES ON LYCOPENE USING EFSA-VALID MARKERS OF OXIDATIVE DAMAGE PREVENTION

The European regulatory process gives much relevance to the establishment, on the basis of a hierarchy of scientific methodologies, of cause-and-effect relationships between food intake and desired health outcomes. This hierarchy, as defined by EFSA, places human intervention studies and, particularly, randomized controlled trials (RCTs) at the top.

Considering all previous information, we performed a search on international scientific publications PubMed-NCBI (http://www.ncbi.nlm.Nih.gov/PubMed) of the last five years using the key words "lycopene" and "antioxidants" with a result of 226 studies. Taking into account the type of study, 111 are review articles and 109 correspond to clinical trials, of which seven are meta-analysis (the types of studies with the highest level of scientific evidence). The most recent meta-analyses were carried out in 2013 and 2015 by Chen and his team on lycopene/tomato consumption and the risk of prostate cancer. Yang et al. (2013) showed the role of tomato products and lycopene in the prevention of gastric cancer. Valero and his research group (2011) published a meta-analysis on the role of lycopene in type-2 diabetes mellitus. The protective effect of lycopene on serum cholesterol and blood pressure was shown by meta-analyses of intervention trials (Ried and Fakler 2011).

Now, in Table 10.4 we show those studies carried out on the antioxidant activity of lycopene (protection against oxidative damage) in which some of the valid oxidative damage prevention markers required by EFSA are used and the limitations we have detected.

TABLE 10.4
Studies on Lycopene Using Valid Markers of Oxidative Damage Prevention

Valid Marker	References	Limitations
Risk Factor: Oxidative Lipid Damage (Lipid Peroxidation)		
Levels of F2-isoprostanes in 24-h urine samples by CG-MS (immunoassay is not very specific). Level of oxidized LDL in blood by immunological methods (antibodies with appropriate specificity). Hydroperoxides of phosphatidylcholine (PCOOH) in blood or tissue by HPLC. Levels of malondialdehyde MDA or blood lipid peroxidase levels in blood or tissues can only be used as evidence of support (F2-isoprostanes and oxidized LDL *in vivo*) by HPLC.	(Visioli et al. 2003, pp. 201–6) (Silaste et al. 2007, pp. 1251–8)	Uncontrolled intervention trials. Sample size n = 12. Only women; dose 8 mg lycopene/d. Diet rich in tomato. Level of oxidized LDL in blood not by immunological methods. Sample size n = 21; dose 30 mg lycopene/d. Not well-defined control. 400 ml of tomato juice and 30 mg of ketchup per day.
Risk Factor: Oxidative Damage to Proteins		
Oxidation products of proteins and aa by HPLC-MS. DNPH and ELISA only useful as a complement in samples of blood or target tissue (e.g., skin).	(Rizwan et al. 2011, pp. 154–62)	Sample size n = 20; dose 16 mg lycopene/d of 55 g of tomato paste. No HPLC-mass.

(Continued)

TABLE 10.4 (CONTINUED)

Studies on Lycopene Using Valid Markers of Oxidative Damage Prevention

Valid Marker	References	Limitations
Risk Factor: Oxidative Damage to DNA		
Modifications of the kite test *in vivo*, which allow the detection of oxidized DNA bases.	(Riso et al. 1999, pp. 712–18) (Riso et al. 2006, pp. 2563–6) (Kim et al. 2011, pp. 189–95) (Devaraj et al. 2008)	Sample size n = 10; dose 16.5 mg lycopene/d of tomato purée. No modified comet test. Sample size n = 26. It is not just lycopene, it is ly-comato (lycopene, beta carotene, and tocopherols). LycoRed Natural Products Industries Ltd., Beer-Sheva, Israel. Sample size n = 26. It is not just lycopene, it is ly-comato (lycopene, beta carotene, and tocopherols). Sample size n = 21; dose 30 mg lycopene/d. Supplementation 6.5/15 and 30 mg. Roche DMS.
Quantification of 8-hydroxy-2-deoxyguanosine (8-OHdG) in blood, tissues, or urine.	(Porrini et al. 2005, pp. 93–9) (Devaraj et al. 2008)	Sample size n = 26 It is not just lycopene Ly-comato (lycopene, beta carotene, tocopherols.) LycoRed Natural Products Industries Ltd., Beer-Sheva, Israel. Sample size n = 21; dose 30 mg lycopene/d. Supplementation 6.5/15 and 30 mg. Roche DMS.

It should be noted that none of the selected studies fully complies with the requirements of EFSA to support the claims of antioxidant properties of lycopene, its antioxidant capacity, and lycopene as an antioxidant defence system. This is because there is no well-defined control study, or the population under study is not a healthy population but presents certain pathologies; or, in spite of using the appropriate markers, the analytical techniques used for their quantification are not those required by EFSA.

10.7 CONCLUSIONS

Regarding the lack of human studies, the NDA panel considers that evidence derived from animal or *in vitro* studies may support the reported effect in humans only if data are provided on the mechanisms by which the food/constituent could exert the alleged effect, and the biological plausibility of the specific claim is documented.

According to the EFSA, antioxidant properties of food (evaluated *in vitro*), and changes in total antioxidant capacity of plasma (measured *in vivo* with methods such as TRAP, TEAC, FRAP, ORAC, or FOX), do not predict the role of the food/constituent in the protection of body cells and molecules such as DNA, proteins, and lipids from oxidative damage *in vivo*, and therefore are not adequate measures for the scientific basis of the claimed effect.

For claims related to the "antioxidant defence system" and the evaluation of the effects of food/constituents on enzymes and compounds (e.g., glutathione) that form part of the endogenous antioxidant system of the human body, EFSA considers that some of the essential vitamins and minerals have a role in the function of the enzymes belonging to the human antioxidant network that protects cells and molecules from oxidative damage. In the context of an adequate supply of these essential vitamins and minerals, a specific induction of antioxidant enzymes such as superoxide dismutase (SOD), catalase, glutathione peroxidase (GSH-Px), and hemoxygenase, or limiting the decrease in glutathione, indicates a biological response to oxidative stress of any origin. EFSA considers a beneficial physiological effect only if such changes provide (additional) protection for cells and molecules against oxidative damage and are demonstrated *in vivo* in humans. Thus, the induction of antioxidant enzymes cannot be used alone as evidence of the claims relating to the "antioxidant defence system" for constituents of non-essential foods (as in the case of lycopene).

Although many studies link tomato consumption and lycopene with the reduction of some risk factors of disease, so far, according to the health claims regulation, there is only one claim authorized under Article 13(5): "reduction of platelet aggregation."

None of the reviewed studies fully complies with the requirements of EFSA to support the claims of antioxidant properties of lycopene, its antioxidant capacity and lycopene as an antioxidant defence system. It would be necessary to conduct well-controlled human intervention studies where the population included healthy individuals and the markers considered valid by the EFSA would be used and quantified according to the appropriate analytical methodology.

ACKNOWLEDGMENTS

The authors are grateful to the OTRI UCM Tomato Foundation (224-2016) project.

REFERENCES

Berliner J. A. and Heinecke J. W. (1996). The role of oxidized lipoproteins in atherogenesis. *Free Radical Biology and Medicine* 20:707–27.

Cámara, M. and Sánchez-Mata, M. C. (2006). Tomatoes, Lycopene and Human Health. In *Lycopene Analysis in Foods*, ed. Rao V., Caledonian Press: Barcelona, Spain, 260 (7):9–62.

Cámara, M., Fernández-Ruiz, V., Fernández-Redondo, D., Sánchez-Mata, M., Cámara, R. M. and Gervás, C. (2012). EFSA scientific requirements related to lycopene as antioxidant prevention of oxidative damage and cardiovascular health claims. En 13th ISHS Symposium on the Processing Tomato, 40–1.

Cámara, M., Sánchez-Mata, M. C., Fernández-Ruiz, V., Cámara, M. R., Manzoor, S. and Cáceres, J. O. (2013). Lycopene: A Review of Chemical and Biological Activity Related to Beneficial Health Effects. In *En Studies in Natural Products Chemistry*, ed. Atta-ur-Rahman, Elsevier B.V.

Cámara, M., Fernández-Ruiz, V., Fernández Redondo, D., Sánchez-Mata, M. C., Cámara, R. M., and Gervás, C. (2015). EFSA scientific requirements related to lycopene as antioxidant prevention of oxidative damage and cardiovascular health claims. *Acta Horticulturae, International Society for Horticultural Sciences*. Eds Cámara, M, Battilani, A, Colvine, S., Leuven (Bélgica). *Acta Horticulturae* 1081:303–7.

Chen, P., Zhang, W., Wang, X., Zhao, K., Negi, D. S., Zhuo, L., and Zhang, X. (2015). Lycopene and risk of prostate cancer: A systematic review and meta-analysis. *Medicine* 94(33):e1260.

Commission Regulation (EU) No 432/2012 of 16 May 2012 establishing a list of permitted health claims made on foods, other than those referring to the reduction of disease risk and to children's development and health.

Commission Regulation (EU) No 1048/2012 of 8 November 2012 on the authorization of a health claim made on foods and referring to the reduction of disease risk.

Devaraj, S., Mathur, S., Basu, A., Aung, H. H., Vasu, V. T., Meyers, S., and Jialal, I. (2008). A dose-response study on the effects of purified lycopene supplementation on biomarkers of oxidative stress. *Journal of the American College of Nutrition* 27(2):267–273.

EFSA Journal (2009a). Scientific Opinion. Water-soluble tomato concentrate (WSTC I and II) and platelet aggregation Scientific substantiation of a health claim related to water-soluble tomato concentrate (WSTC I and II) and platelet aggregation pursuant to Article 13 (5) of Regulation (EC) No 1924/2006. *EFSA Journal* 1101:1–15.

EFSA Journal (2009b). Scientific Opinion. Lycopene-whey complex (bioavailable lycopene) and risk of atherosclerotic plaques Scientific substantiation of a health claim related to Lycopene-whey complex (bioavailable lycopene) and reduction of the risk of atherosclerotic plaques pursuant to Article 14 of Regulation (EC) No 1924/2006. *EFSA Journal* 1179:1–10.

EFSA Journal (2011). Scientific Opinion on the substantiation of health claims related to lycopene and protection of DNA, proteins and lipids from oxidative damage (ID 1608, 1609, 1611, 1662, 1663, 1664, 1899, 1942, 2081, 2082, 2142, 2374), protection of the skin from UV-induced (including photo-oxidative) damage (ID 1259, 1607, 1665, 2143, 2262, 2373), contribution to normal cardiac function (ID 1610, 2372), and maintenance of normal vision (ID 1827) pursuant to Article 13(1) of Regulation (EC) No 1924/2006. *EFSA Journal* 9(4):2031.

Griffiths, H. R., Moller, L., Bartosz, G., Bast, A., Bertoni-Freddari, C., Collins, A., Cooke, M., Coolen, S., Haenen, G., Hoberg, A. M., Loft, S., Lunec, J., Olinski, R., Parry, J., Pompella, A., Poulsen, H., Verhagen, H., and Astley, S. B. (2002). Biomarkers. *Molecular Aspects of Medicine* 23:101–208.

Halliwell, B. (1987). Oxidants and human disease: Some new concepts. *Federation of American Societies for Experimental Biology* 1:358.

ILSI Europe (Internacional Life Science Intitute). (1999). FOFOSE, Scientific concepts of functional foods in Europe. Consensus Documents. *Journal of Nutrition* 81: 15–27S.

Jinyao, C. H. E. N., Yang, S. O. N. G., and Zhang, L. (2013). Lycopene/tomato consumption and the risk of prostate cancer: A systematic review and meta-analysis of prospective studies. *Journal of Nutritional Science and Vitaminology* 59(3):213–23.

Kim, J. Y., Paik, J. K., Kim, O. Y., Park, H. W., Lee, J. H., Jang, Y., and Lee, J. H. (2011). Effects of lycopene supplementation on oxidative stress and markers of endothelial function in healthy men. *Atherosclerosis* 215(1):189–95.

Knasmüller, S., Nersesyan, A., Misik, M., Gerner, C., Mikulits, W., Ehrlich, V., Hoelzl, C., Szakmary, A., and Wagner, K. H. (2008). Use of conventional and -omics based methods for health claims of dietary antioxidants: A critical overview. *British Journal of Nutrition* 99 E Suppl 1:ES3–52.

Kong, K. W., Khoo, H. E., Prasad, K. N., Ismail, A., Tan, C. P., and Rajab, N. F. (2010). Revealing the power of the natural red pigment lycopene. *Molecules* 15:959–87.

Lampe, J. W. (1999). Health effects of vegetables and fruits: Assessing the mechanisms of action in human experiments studies. *American Jounal of Clininical Nutrition* 70:475S–90S.

Mendis, S., Puska, P., and Norrving, B. (2011). Global atlas on cardiovascular disease prevention and control. World Health Organization. Available at http://apps.who.int/iris/handle/10665/44701 (Accessed 29/08/14).

Mensink, R. P., Aro, A., Den Hond, E., German, J. B., Griffin, B. A., ten Meer, H. U., Mutanen, M., Pannemans, D., and Stahl, W. (2003). PASSCLAIM - Diet-related cardiovascular disease. *European Journal of Nutrition* 42 Suppl 1:I6–27.

Miller, E. R., Erlinger, T. P., Sacks, F. M., Svetkey, L. P., Charleston, J., Lin, P., and Appel, L. J. (2005). A dietary pattern that lowers oxidative stress increases antibodies to oxidized LDL: Results from a randomized controlled feeding study. *Atherosclerosis* 183:175–82.

Porrini, M., Riso, P., Brusamolino, A., Berti, C., Guarnieri, S., and Visioli, F. (2005). Daily intake of a formulated tomato drink affects carotenoid plasma and lymphocyte concentrations and improves cellular antioxidant protection. *British Journal of Nutrition* 93(1):93–9.

Rao, A. V. (2006). *Tomatoes, Lycopene & Human Health: Preventing Chronic Diseases*, Caledonian Science Press.

Regulation (EC) No 1924/2006 of the European Parliament and of the Council of 20 December 2006 on nutrition and health claims made on foods.

Ried, K., and Fakler, P. (2011). Protective effect of lycopene on serum cholesterol and blood pressure: Meta-analyses of intervention trials. *Maturitas* 68(4):299–310.

Riso, P., Pinder, A., Santangelo, A., and Porrini, M. (1999). Does tomato consumption effectively increase the resistance of lymphocyte DNA to oxidative damage? *The American Journal of Clinical Nutrition* 69(4):712–8.

Riso, P., Visioli, F., Grande, S., Guarnieri, S., Gardana, C., Simonetti, P., and Porrini, M. (2006). Effect of a tomato-based drink on markers of inflammation, immunomodulation, and oxidative stress. *Journal of Agricultural and Food Chemistry* 54(7):2563–6.

Rizwan, M., Rodriguez-Blanco, I., Harbottle, A., Birch-Machin, M. A., Watson, R. E. B., and Rhodes, L. E. (2011). Tomato paste rich in lycopene protects against cutaneous photodamage in humans in vivo: A randomized controlled trial. *British Journal of Dermatology* 164(1):154–2.

Seelert, K. (1992). Antioxidants in the prevention of atherosclerosis and coronary heart disease. *Internist Prax* 32:191–9.

Silaste, M. L., Alfthan, G., Aro, A., Kesäniemi, Y. A., and Hörkkö, S. (2007). Tomato juice decreases LDL cholesterol levels and increases LDL resistance to oxidation. *British Journal of Nutrition* 98(6):1251–8.

Yang, T., Yang, X., Wang, X., Wang, Y., and Song, Z. (2013). The role of tomato products and lycopene in the prevention of gastric cancer: A meta-analysis of epidemiologic studies. *Medical hypotheses* 80(4):383–8.

Valero, M. A., Vidal, A., Burgos, R., Calvo, F. L., Martínez, C., Luengo, L. M., and Cuerda, C. (2011). Meta-analysis on the role of lycopene in type 2 diabetes mellitus. *Nutrición Hospitalaria* 26(6):1236–41.

Visioli, F., Riso, P., Grande, S., Galli, C., and Porrini, M. (2003). Protective activity of tomato products on in vivo markers of lipid oxidation. *European Journal of Nutrition* 42(4):201–6.

Index

Page numbers followed by f and t indicate figures and tables, respectively.